Birds - Conservation, Research and Ecology

Edited by Heimo Mikkola

Published in London, United Kingdom

Birds - Conservation, Research and Ecology
http://dx.doi.org/10.5772/intechopen.1000275
Edited by Heimo Mikkola

Contributors
François Sueur, Ashraf S. Alias, Garima Singh, Sachin Kumar, Kalpana Chaudhary, Gaurav Sharma, Luis Fernando Basanta Reyes, Heimo Mikkola, Anita Mikkola, Ally K. Nkwabi, Pius Y. Kavana, Arie M. Rijke, Julio Luis Erwin Illanes Herrero, Mahalakshmi Sundarapandian, Priya Raman

First published in London, United Kingdom, 2023 by IntechOpen
IntechOpen is the global imprint of INTECHOPEN LIMITED, registered in England and Wales, registration number: 11086078, 5 Princes Gate Court, London, SW7 2QJ, United Kingdom

British Library Cataloguing-in-Publication Data
A catalogue record for this book is available from the British Library

Additional hard and PDF copies can be obtained from orders@intechopen.com

Birds - Conservation, Research and Ecology
Edited by Heimo Mikkola
p. cm.
Print ISBN 978-1-83769-619-2
Online ISBN 978-1-83769-618-5
eBook (PDF) ISBN 978-1-83769-620-8

Meet the editor

Heimo Mikkola, Ph.D., is an adjunct professor at the University of Kuopio, Finland (now the University of Eastern Finland). He has been a birdwatcher since the age of eleven and has visited 137 countries of the world (46 of those in Africa) and observed various birds in all of them. He has worked for international organizations like the African Development Bank, European Union, United Nations System, and the World Bank for more than 35 years. Dr. Mikkola has published almost 700 papers and books, mainly on birds and other animals, and has edited eleven books.

Contents

Preface

Birdwatching is the second most popular hobby in the Western world. There has also been a large increase in the use of data gathered by amateur birdwatchers. Cunning researchers have tapped backyard birders by asking them to complete various kinds of bird surveys [1]. Christmas or Winter Bird Count Weeks, the Bird Nest Box Surveys, Global Big Day, and Breeding Bird Atlases are all great ways to apply one's knowledge to collect data that can be used for broadscale, long-term trend analysis. The Cornell Lab of Ornithology in the New World has a project called eBird (www.ebird.org) and there are BirdLife International equals elsewhere (f.i. www.artobservasjoner.no; www.artportalen.se; www.putni.lv; www.tiira.fi, etc.). These are Internet tools in which every birdwatcher can easily and quickly enter any bird into a database that can be accessed by researchers and the general public. Birders can also access their bird observations whenever they want, allowing them an easy way to look at their records in different ways. A system like eBird makes it possible to answer your personal questions about what bird species you recorded and when and where you saw them [2]. In eBird, every bird sighting represents a little data point in the big puzzle, in which we try to understand the changes in bird distribution and populations [2].

However, citizen data have some possible biases, including uneven geographical distributions of birdwatchers and habitat-biased observation efforts. Birdwatcher motivations may also cause some unwanted influences as some observers report their bird records stemming from general outdoor activities while other birders only visit specific areas in search of rare species [3]. Therefore, citizen data must be evaluated with caution and possible biases need to be identified [4].

We can safely use eBird and BirdLife International data to predict patterns of occurrence, such as irruptions [5, 6]. In Norway, the value of citizen data has been evaluated when studying owl irruptions and it was concluded that citizen and survey data from the same area gave similar results [7]. In addition, citizen data have the advantage that large amounts of information are collected over wide areas [7]. However, often it is necessary to remove the overlapping observations most likely concerning the same bird reported by two or more observers. In countries where the species has to be reported to the respective national rarities committee, the citizen data can be compared with the officially accepted records and adjusted for those that remained undocumented [8].

In some studies, concern has been raised about people's ability to identify bird vocalizations 100 percent correctly, especially the often difficult voices. Therefore it is a remarkable improvement that new mobile applications have been created to identify singing species with your cellular phone. The world's leading app for birds is the American Merlin Bird ID, which helps you identify birds you see and hear. Merlin is powered by eBird, the world's largest database of bird sounds and sightings [9]. In Finland, one of those applications is from the University of Jyväskylä and it is said to identify 150 birds [9]. Finnish birdwatchers have concluded that artificial intelligence is not replacing well-seasoned birdwatchers [10] but will surely help those less experienced wanting to learn to identify bird voices. In America, users have complained that sometimes the app takes more than one minute to load, and by then the bird they wanted to identify is gone [9].

We should always remember that the goal of birding with a purpose is to protect the birds themselves. The more people get excited about birds, the more people are likely to protect birds. This book is devoted to bird conservation. Humans are just getting back to normal after a devastating COVID-19 pandemic and now birds at least in Europe are suffering from highly pathogenic H5N1 avian influenza. In this book, we have the first warning from France where the virus has been killing gulls. The virus is also present in Finland, but now the disease has mutated and affects fur farm animals. Bird–mammal infection is an alarming sign and people are frightened that the virus will mutate again in mammals and spread to human populations as COVID-19 did. Personal hand hygiene and avoiding any direct contact with sick or dead birds and mammals is a good precaution.

The book includes nine chapters dealing with agricultural influence, anthropogenic noise, avian influenza, cholinesterase activity, ecology, feather structure, histology, and conservation of birds all over the globe. I want to express my thanks to the Author Service Manager and team leader Ivana Barać at IntechOpen for her active and skillful cooperation during the preparation and publication of this book.

Heimo Mikkola
University of Eastern Finland,
Kuopio, Finland

References

[1] O'Connor M. Geeks of a feather. New Scientist 2005; May 7, p. 21

[2] Lovitch D. How to Be a Better Birder. Princeton, NJ: Princeton University Press; 2012

[3] Tulloch ATT, Szabo JK. A behavioural ecology approach to understand volunteer surveying for citizen sciences datasets. Emu – Austral Ornithology. 2012;**112**:313-325

[4] Dickinson JL, Shirk J, Bonter DN, Bonney R, Crain RL, Martin J, et al. The current state of citizen science as a tool for ecological research and public engagement. Frontiers in Ecology and the Environment. 2012;**10**:291-297

[5] Dale S. Density, numbers and probable origin of Northern Hawk-Owls Surnia ulula in southern Norway during the 2016 irruption. Ornis Norvegica. 2017;**40**:1-13

[6] Dale S, Sonerud GA. Hawk owl irruptions: spatial and temporal variation in rodent abundance drive push and pull dynamics. Oecologia. 2022; DOI: 10.1007/s00442-022-05283-9

[7] Dale S. Relative magnitude of two Northern Hawk-Owl *Surnia ulula* irruptions to southern Norway: comparison of citizen data and survey data. Ornis Norvegica. 2022;**45**:1-9. DOI: 10.15845/on.v45i0.3425

[8] Mikkola H, Rajala E, Väli Ü, Keišs O, Jusus V, Kwieciński Z, et al. The northern hawk owl Surnia ulula invasions in Europe. Raptor Journal. 2023;**17**:1-14. DOI: 10.2478/srj-2023-0001

[9] Cornell Lab of Ornithology. Merlin Bird ID by Cornell Lab. Apps on Google Play. [Accessed: July 5, 2023]

[10] Arkiomaa A. Mullistaako tekoäly tietomme luonnosta? Linnut-Vuosikirja 2022:3. published 2023

Chapter 1

Impact of Highly Pathogenic H5N1 Avian Influenza on Some Gull Populations on the Coast of Hauts-de-France (France)

François Sueur

Abstract

In recent years, highly pathogenic H5N1 avian influenza has occasionally plagued wild bird populations. For the first time in 2022, it was particularly virulent on the gull populations of the Hauts-de-France coast (France). This work analyzes the consequences of the virus on the numbers and production of young in three colonies of gulls (Laughing Gull, Mediterranean Gull, and Sandwich Tern) on the coast of Pas-de-Calais and the Somme by comparing with the results of the years previous pandemic-free periods. This work also compares the numbers of roosts of several species of gulls in the Authie-Somme reserve obtained in spring to autumn in 2022 with those of previous years.

Keywords: highly pathogenic H5N1 avian influenza, gull populations, Hauts-de-France, pandemic, gull mortality

1. Introduction

Cases of joint mass mortalities involving humans, domestic mammals, and birds have been known since antiquity (1200 BC) but scientific evidence that it is avian influenza does not appear until the end of the nineteenth century [1].

For many decades, wild birds, especially ducks, have been known to be healthy carriers of various influenza viruses.

The H5N1 strain was detected in a few geese in 1996 in China and then the following year in Hong Kong, where it affected humans and chicken farms [1].

A new emergence of this strain occurs in 2004 in East Asia. It spread the following year to Turkey and Romania and then reached Western Europe in 2006 (including France). Another wave was reported in 2015 with cases recorded, particularly in Africa and North America [2].

The role of migratory birds in the dispersal of H5N1 has been the subject of numerous studies [3–6] as has its impact on poultry farms [7–10], but the one on wild birds remains unknown.

For the first time in 2022, this H5N1 strain was particularly virulent in the gull populations of Hauts-de-France (France). Also, an attempt to measure the influence of this virus on these birds was made. This work analyzes the consequences of the virus on gull mortality, breeding numbers, and the production of young in three colonies (Black-headed Gull *Chroicocephalus ridibundus*, Mediterranean Gull *Ichthyaetus melanocephalus,* and Sandwich Tern *Thalasseus sandvicensis*) on the coasts of Pas-de-Calais and of the Somme (natural region of Marquenterre). It also analyzes the consequences of the resting places of several species of gulls in the Authie-Somme reserve.

2. Methods

Gull corpses were counted at the various sites studied.

The colonies of three localities of Marquenterre (Ornithological Park of Marquenterre in Saint-Quentin-en-Tourmont in the Somme, Conchil-le-Temple, and Groffliers in the Pas-de-Calais) were counted at least once per decade of March to July. No incursion is carried out on the colonies which are observed from a distance (from observation posts in Saint-Quentin-en-Tourmont and Conchil-le-Temple). Whenever possible, breeding adults were distinguished from nonbreeding immature. Behaviors related to reproduction (copulations, transport of materials, and incubating nests) were also noted. As soon as they appear, the chicks are also counted. The species concerned are the Black-headed Gull, the Mediterranean Gull, and the Sandwich Tern.

In the Authie-Somme reserve, gull roosts are counted or estimated (in the case of numbers of several thousand individuals of a single species) throughout the annual cycle. Rare taxa (Bonaparte's Gull *Chroicocephalus philadelphia*, Glaucous Gull *Larus hyperboreus,* and Caspian Tern *Hydroprogne caspia*), pelagic (Kittiwake Gull *Rissa tridactyla* and Little Gull *Hydrocoloeus minutus*) or with too small numbers (Little Tern *Sternula albifrons*, Arctic Tern *Sterna paradisaea,* and Black Tern *Chlidonias niger*).

In both types of data, the numbers of the years without the emergence of the H5N1 virus (2018 to 2021) are compared to those of the year with the virus. Due to COVID-related traffic restrictions, data from mid-March to early May 2020 was sometimes not obtained.

3. Results

The first gull victim of H5N1 is an immature Herring Gull *Larus argentatus* discovered on May 18, 2022 on the coast of the town of Saint-Quentin-en-Tourmont. In 3 weeks, more than 800 birds (probably around a thousand), mainly gulls, were found dead in the study area, particularly in the Authie-Somme Reserve and in the Somme Bay National Nature Reserve bordering on the first.

Of the three colonies studied, it was mainly the corpses of Black-headed Gulls (**Figure 1**) that were discovered (179 for 6 Mediterranean Gulls and 10 Sandwich Terns).

During the study period, the Black-headed Gull colony at Conchil-le-Temple is expanding with numbers at the end of April 2022, ten times higher than those of 2018 and more than three times those of 2019 and 2021 (**Figure 2**). These numbers fall in May to reach a minimum at the end of June, slightly lower than those of 2021 during the same period but nevertheless remain significantly higher than those of the first three years studied.

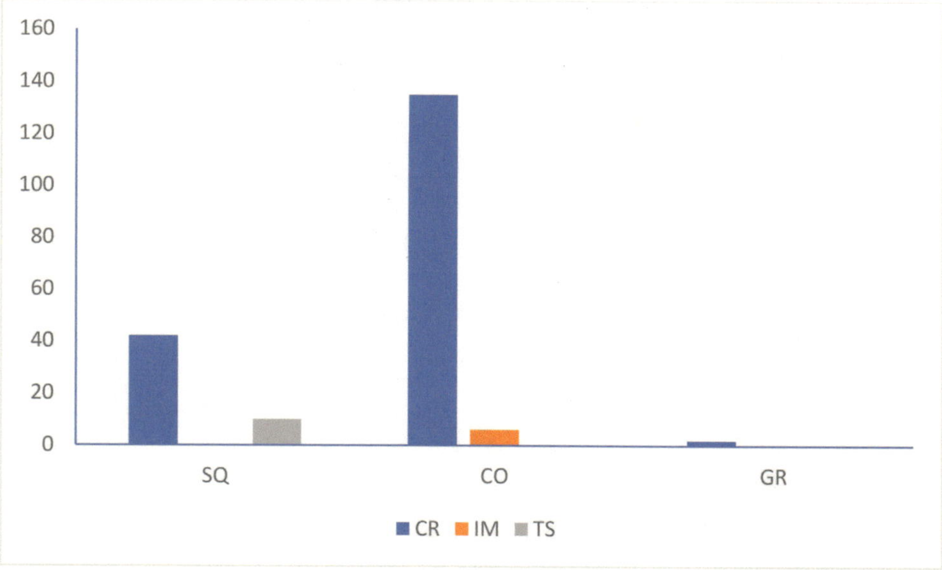

Figure 1.
Number of black-headed Gull Chroicocephalus ridibundus *(CR), Mediterranean Gull* Ichthyaetus melanocephalus *(IM), and Sandwich tern* Thalasseus sandvicensis *(TS) corpses discovered in the three colonies of Saint-Quentin-en-Tourmont (SQ), Conchil-le-Temple (CO), and Groffliers (GR).*

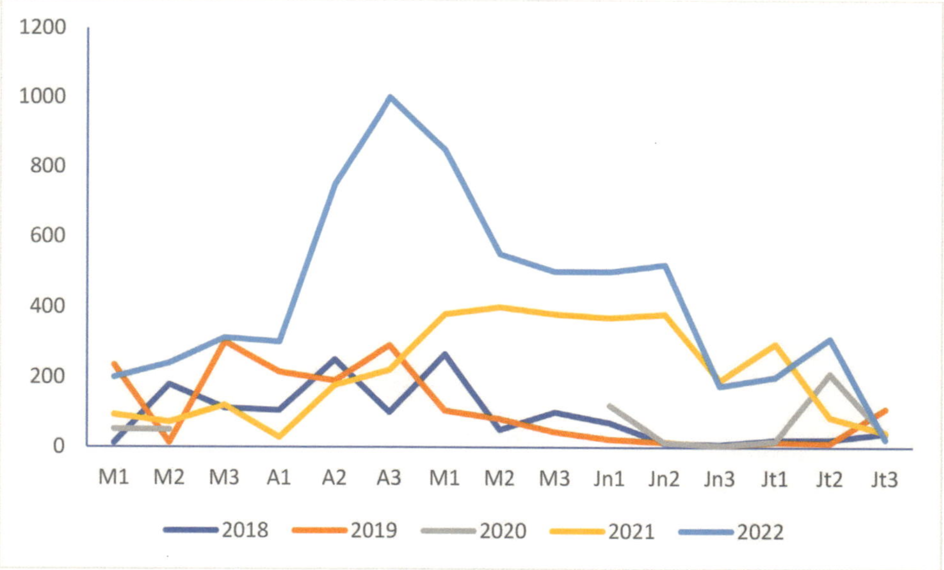

Figure 2.
Evolution of the numbers of the black-headed Gull Chroicocephalus ridibundus *colony of Conchil-le-Temple.*

The situation of the colony of Groffliers is different since the maximum number in 2022 is between the minimum recorded in 2018 and the maximums of 2019, 2020, and 2021 (**Figure 3**). However, they show an earlier (early May) and faster fall than those of these 4 years.

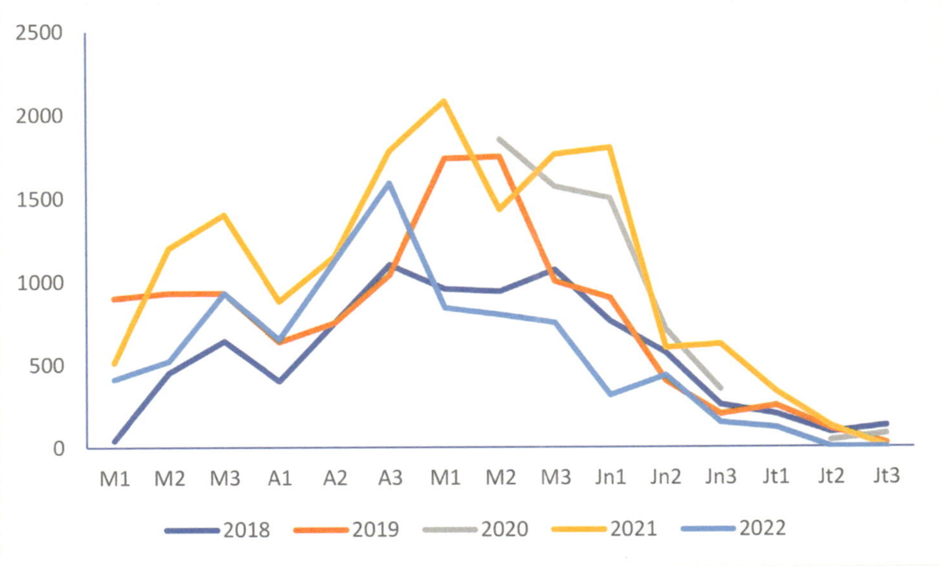

Figure 3.
Evolution of the numbers of the black-headed Gull Chroicocephalus ridibundus *colony of Groffliers.*

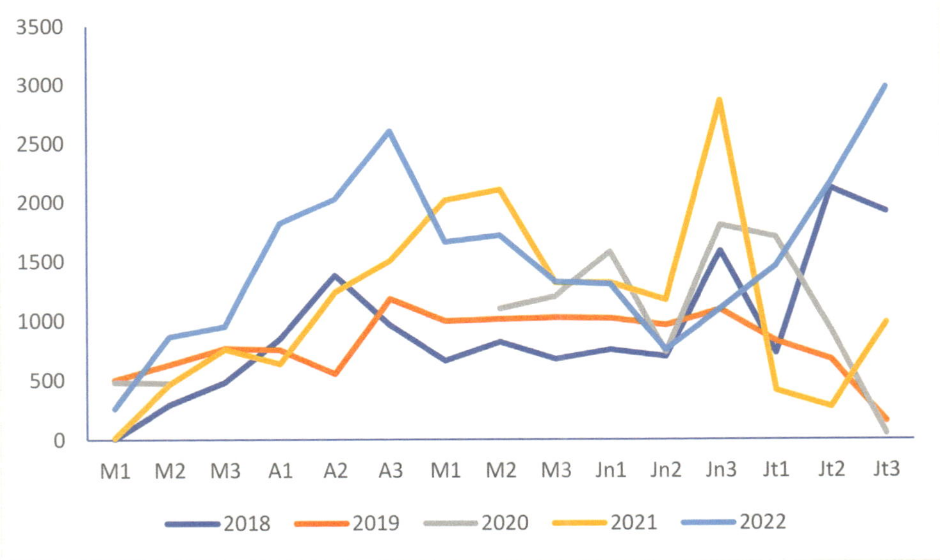

Figure 4.
Evolution of the numbers of the black-headed Gull Chroicocephalus ridibundus *colony of Saint-Quentin-en-Tourmont.*

The situation of the colony of Saint-Quentin (**Figure 4**) is like that of Conchil-le-Temple since the numbers at the end of April are significantly higher than those of the 3 years for which data are available during this decade (2018, 2019, and 2021). They drop to a minimum in mid-June, barely higher than those of 2018 and 2020 but lower than those of 2019 and 2021. These 2022 numbers increase again from the end of June.

The numbers of the Mediterranean Gull colony in Saint-Quentin-en-Tourmont are high at the beginning of April and higher than those of the years 2018, 2019, and 2021 and of the same order of magnitude as those of the first two decades of May 2021, but they drop sharply from the end of May, remain very low until mid-June before increasing until the beginning of July (**Figure 5**).

The Sandwich Tern colony in Saint-Quentin-en-Tourmont has been expanding since 2019. Its numbers at the beginning of May 2022 seem to confirm this fact since they are a little less than five times higher than those of 2021 (**Figure 6**). They show a very clear decrease from the following decade to be at the level of those of 2020 and 2021. This is even more significant with numbers lower than those of the three previous years in July.

In the Black-headed Gull, the production of young in 2022 is lower than that of the four previous years in the colonies of Saint-Quentin-en-Tourmont and Groffliers while on that of Conchil-le-Temple it is only lower than that of 2021 (**Figure 7**).

In the Mediterranean Gull, the production of young in 2022 is higher than that of the four previous years in the colony of Saint-Quentin-en-Tourmont (**Figure 8**). It is very low on that of Conchil-le-Temple: 1 young noted in 2020 and 2022 but 7 in 2021.

In the Authie-Somme Reserve, the differences in numbers from the years 2018 to 2022 do not appear clearly in the Black-headed Gull (**Figure 9**), Common Gull *Larus canus* (**Figure 11**), Great Black-backed Gull *Larus marinus* (**Figure 12**), Herring Gull (**Figure 13**), Yellow-legged Gull *Larus michahellis* (**Figure 14**), Lesser Black-backed Gull *Larus fuscus* (**Figure 15**), Sandwich Tern (**Figure 16**), and Common Tern *Sterna hirundo* (**Figure 17**). Some species even seem to present locally different phenologies depending on the year. For some also, significant numbers are recorded after the period during which the H5N1 virus raged (May to July): Black-headed Gull (mid-August and mid-September to early October), Common Gull (mid-September to early October),

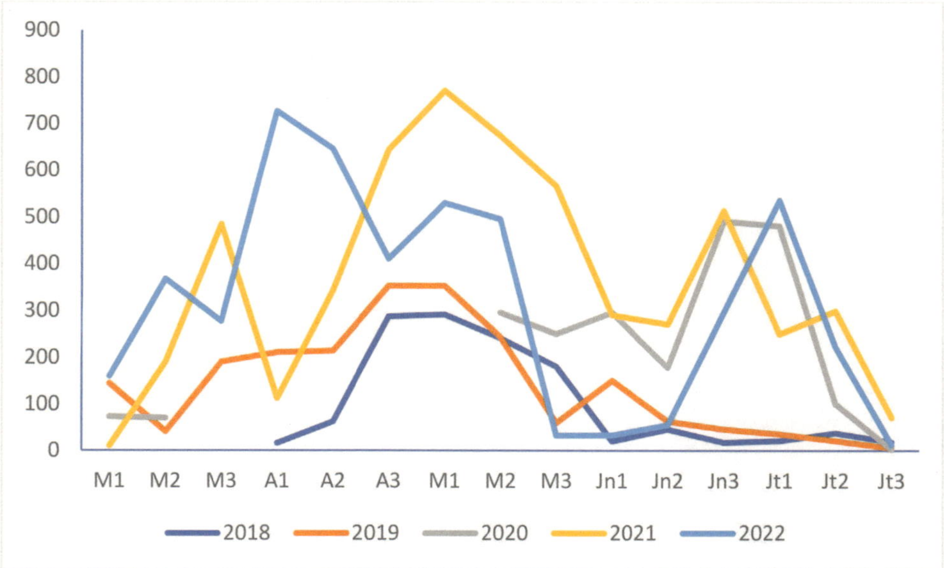

Figure 5.
Evolution of the numbers of the Mediterranean Gull Ichthyaetus melanocephalus *colony of Saint-Quentin-en-Tourmont.*

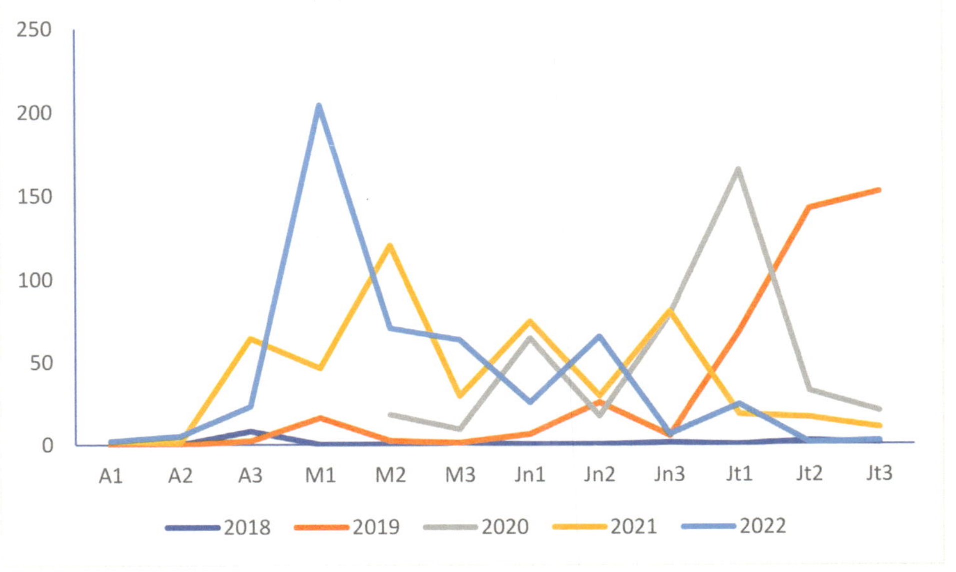

Figure 6.
Evolution of the numbers of the Sandwich tern Thalasseus sandvicensis *colony of SaintQuentin-en-Tourmont.*

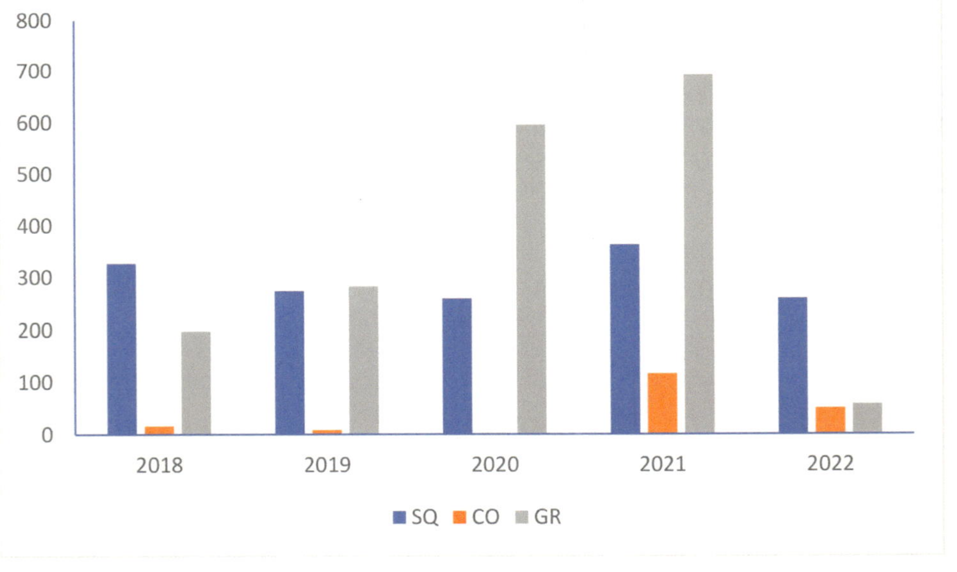

Figure 7.
Number of young black-headed Gull Chroicocephalus ridibundus *observed in the three colonies of Saint-Quentin-en-Tourmont (SQ), Conchil-le-Temple (CO), and Groffliers (GR).*

Great Black-backed Gull (late September), Herring Gull (mid-September to mid-October), Yellow-legged Gull (mid-September), Lesser Black-backed Gull (early to mid-August), Sandwich Tern (early August), and Common Tern (beginning of September).

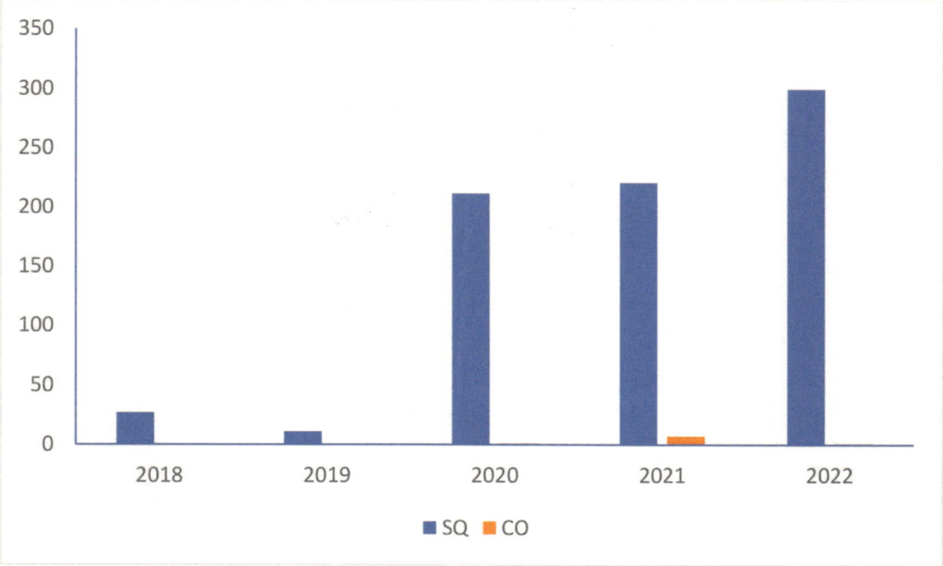

Figure 8.
Number of young Mediterranean Gull Ichthyaetus melanocephalus *observed in the two colonies of Saint-Quentin-en-Tourmont (SQ) and Conchil-le-Temple (CO).*

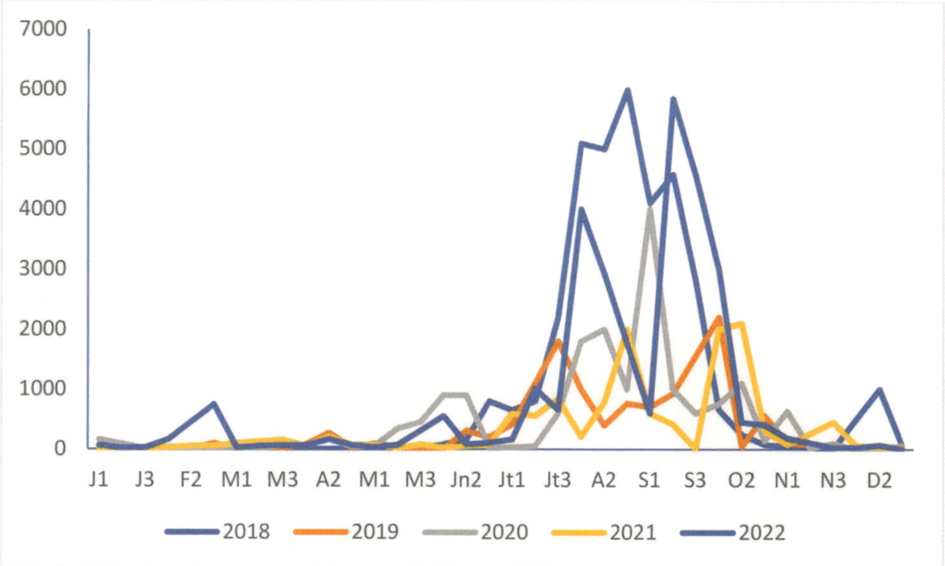

Figure 9.
Black-headed Gull Chroicocephalus ridibundus *numbers in the Authie-Somme reserve during the annual cycle.*

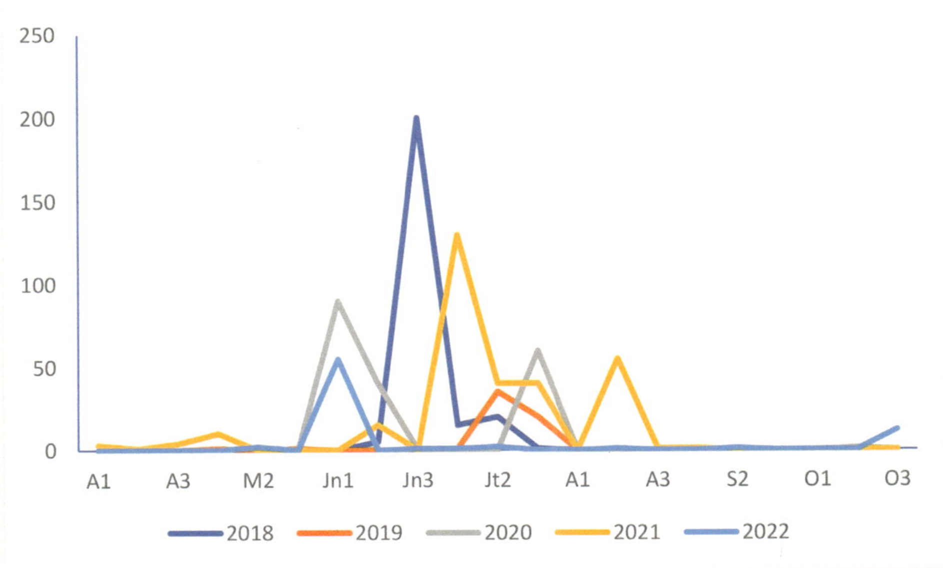

Figure 10.
Mediterranean Gull Ichthyaetus melanocephalus *numbers in the Authie-Somme reserve from April to October.*

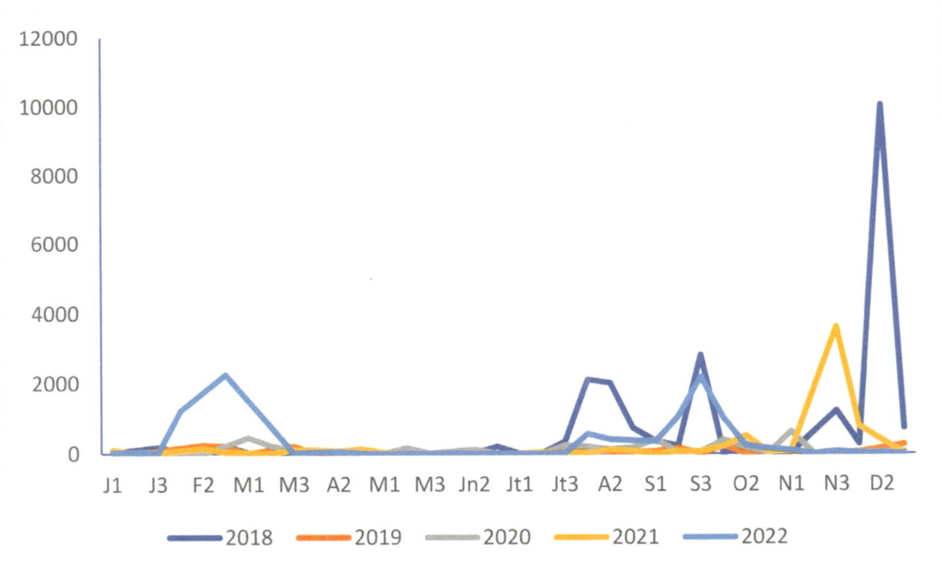

Figure 11.
Common Gull Larus canus *numbers in the Authie-Somme reserve during the annual cycle.*

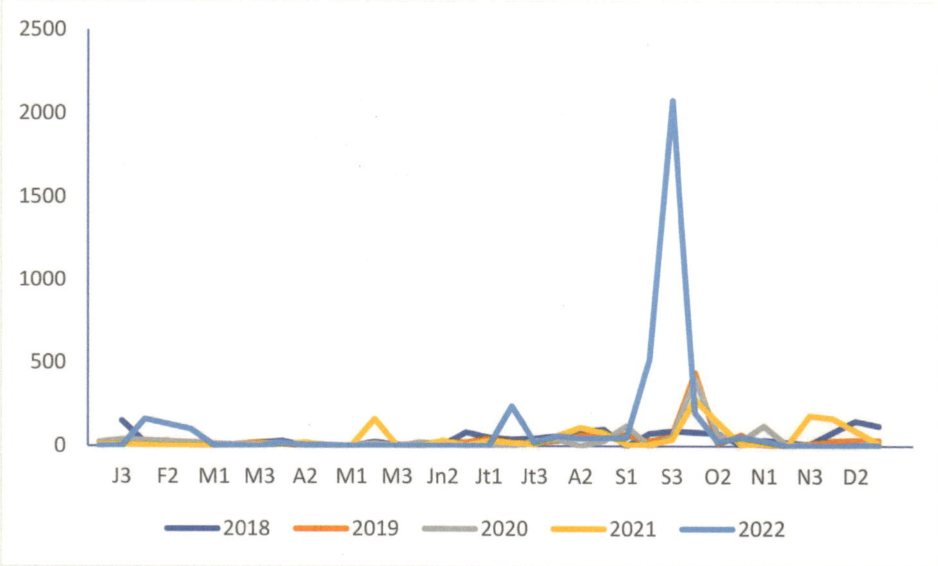

Figure 12.
Great black-backed Gull Larus marinus *numbers in the Authie-Somme reserve during the annual cycle.*

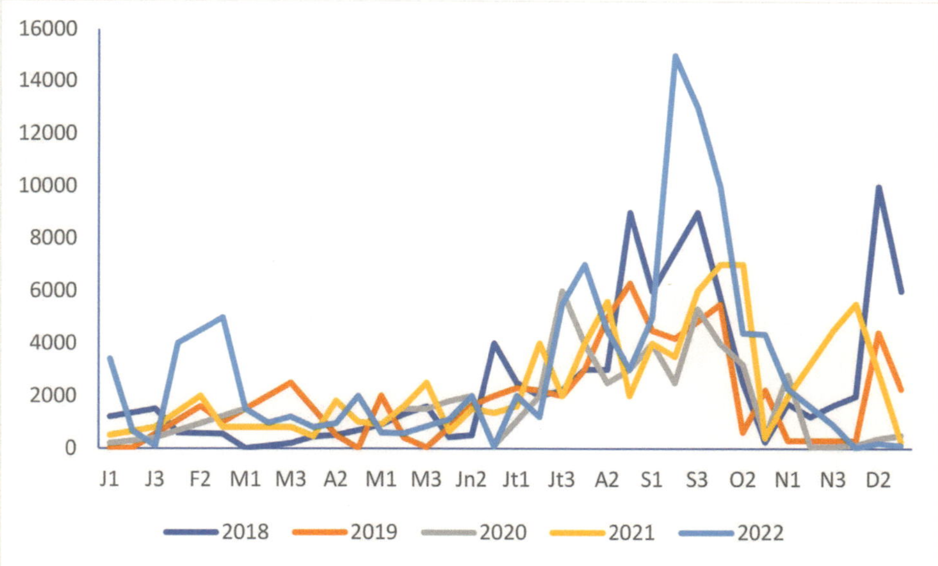

Figure 13.
Herring Gull Larus argentatus *numbers in the Authie-Somme reserve during the annual cycle.*

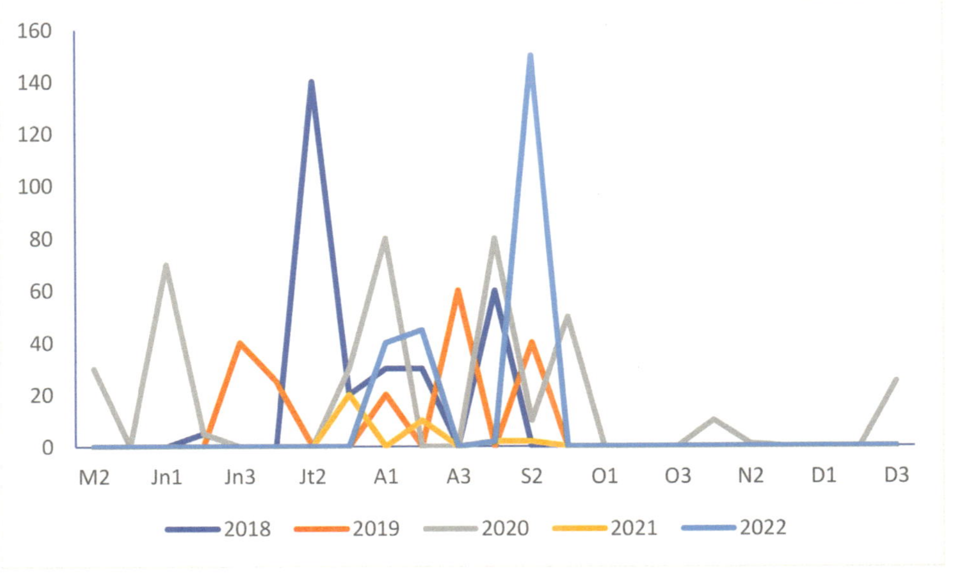

Figure 14.
Yellow-legged Gull Larus michahellis *numbers in the Authie-Somme reserve from mid-may to December.*

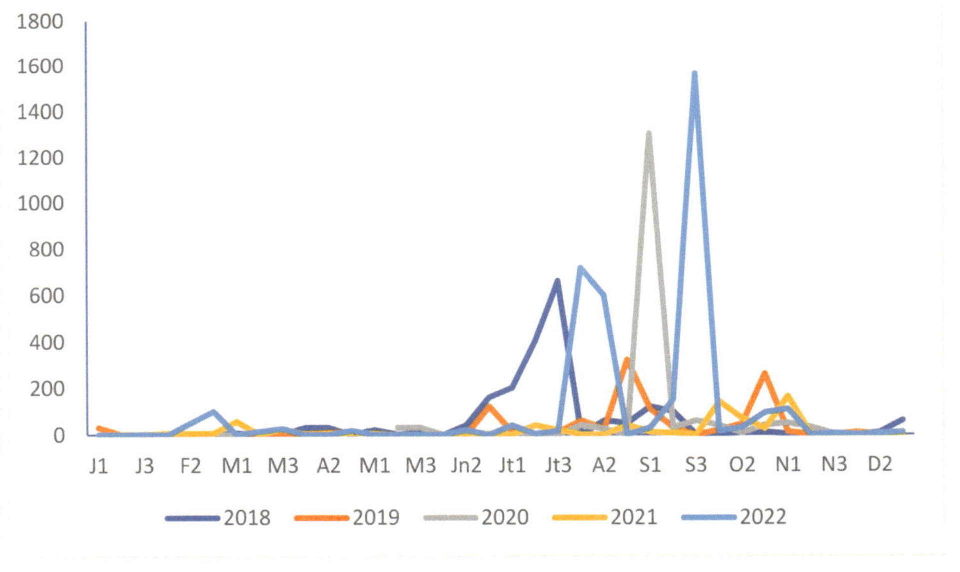

Figure 15.
Lesser black-backed Gull Larus fuscus *numbers in the Authie-Somme reserve during the annual cycle.*

Only the Mediterranean Gull presents significantly lower numbers from mid-June (0 to 2 individuals until mid-October) than in previous years (**Figure 10**).

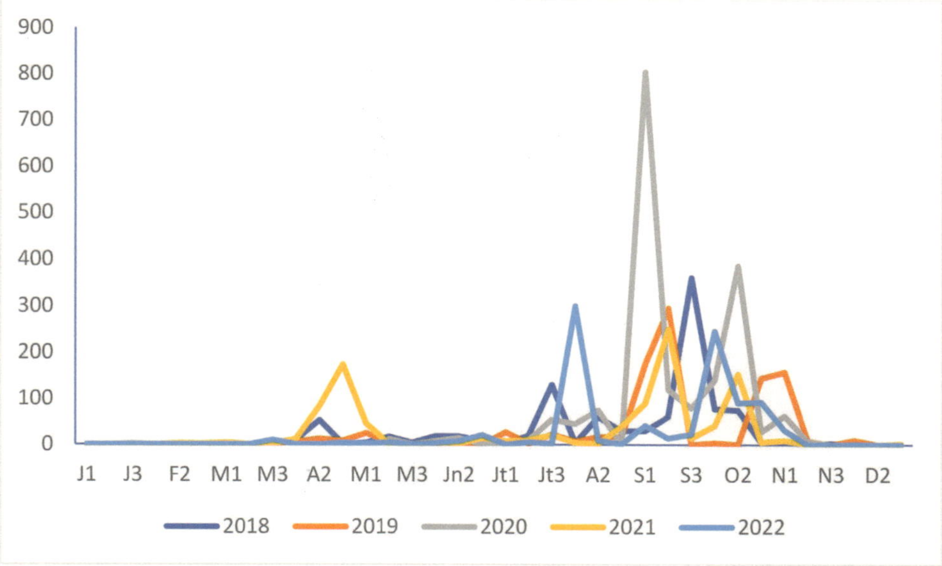

Figure 16.
Sandwich tern Thalasseus sandvicensis *numbers in the Authie-Somme reserve during the annual cycle.*

4. Discussion

The mortality of gulls in the study area is most likely underestimated because collections of dead birds were carried out in the Authie-Somme Reserve and the Somme Bay National Nature Reserve by the municipal services of the communes concerned and by the agents of the OFB (Office Français de la Biodiversité), and on the site of the Marquenterre Ornithological Park and the estuary part of the Somme Bay National Nature Reserve by the staff of these structures. Another source of underestimation is the relatively high vegetation in at least certain sectors of the colonies of Saint-Quentin-en-Tourmont, Conchil-le-Temple, and even more of Groffliers.

The H5N1 virus markedly affected the size of the three colonies and their production of young in the Black-headed Gull. The same is true for the Sandwich Tern in Saint-Quentin-en-Tourmont. However, on this site, the case of the Mediterranean Gull is more complex since the size of the colony is decreasing but the production of young increases in 2022 compared to the four previous years. This fact is probably related to the lower mortality of the Mediterranean Gull compared to that of the other two species (**Figure 1**). To identify the longer-term impact of the virus, it is planned to continue the census of the three colonies studied.

In the Authie-Somme Reserve, where significant mortality was observed from mid-May to June in particular, significant decrease following the emergence of H5N1 could not be demonstrated (except in the Mediterranean Gull) probably due to the massive arrival and prolonged stay of migrants. The very large numbers recorded according to the species between the beginning of August and mid-October are

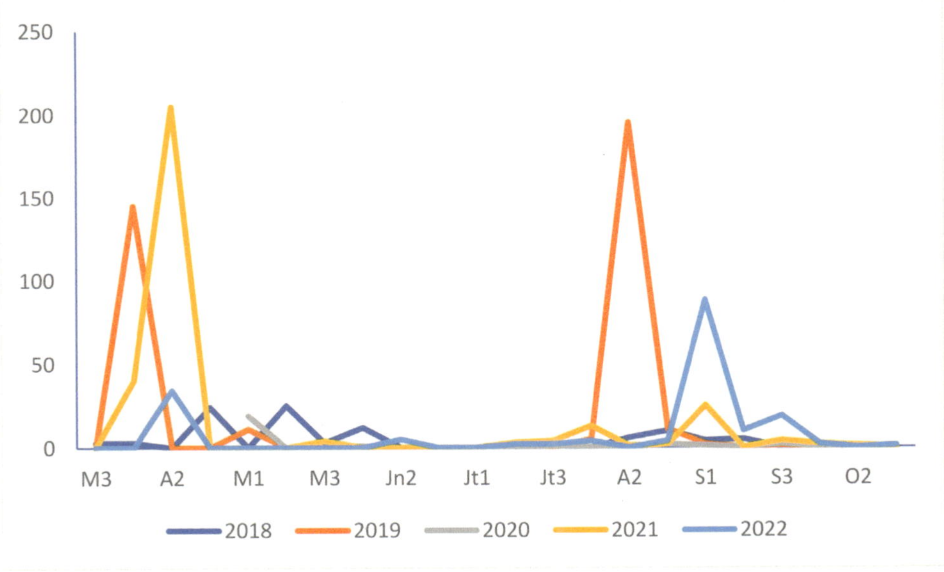

Figure 17.
Numbers of common tern Sterna hirundo *in the Authie-Somme reserve from the end of march to the end of October.*

perhaps caused by a strong increase in the densities of the gulls' prey [11] less consumed by the latter during part of their reproduction period (May and June) [12].

This work constitutes the first approach to the impact of H5N1 on gull populations. Given the dynamics of the colonies and the significant interannual fluctuations in the numbers, it will be necessary to obtain larger annual series to be able to carry out statistical analyses.

Author details

François Sueur
Groupe Ornithologique Picard, Saint-Quentin-en-Tourmont, France

*Address all correspondence to: francois.sueur@wanadoo.fr

References

[1] Blancou J. Histoire de la surveillance et du contrôle des maladies animales transmissibles. Paris: Office international de épizooties; 2000. p. 366

[2] Peyre M, Gaidet N, Caron A, Cappelle J, Tran A, Roger F. Influenza aviaire dans le monde: situation au 31 janvier 2015. Bulletin Epidémiologique – Santé animale, Alimentation. 2015;**6**:10-14

[3] Gauthier-Clerc M, Lebarbenchon C, Thomas F. Recent expansion of highly pathogenic avian influenza H5N1: A critical review. Ibis. 2007;**149**:202-214

[4] Weber TP, Stilianakis NI. Migratory birds, the H5N1 influenza virus and the scientific method. Virology Journal. 2008;**5**:57-59

[5] Caron A, Gaidet N, de Garine-Wichatitsky M, Morand S, Cameron EZ. Evolutionary biology, community ecology and avian influenza research. Infection, Genetics and Evolution. 2009;**9**:298-303

[6] Takekawa JY, Prosser DJ, Newman SH, Muzaffar SB, Hill NJ, Yan B, et al. Victims and vectors: Highly pathogenic avian influenza H5N1 and the ecology of wild birds. Avian Biology Research. 2010;**3**:1-23

[7] Kayali G, Ortiz EJ, Chorazy ML, Gray GC. Evidence of previous avian influenza infection among US Turkey workers. Zoonoses and Public Health. 2009:1-8

[8] Akanbi OB, Taiwo VO. Husbandry practices and outbreak features of natural highly pathogenic avian influenza (H5N1) in Turkey flocks in Nigeria, 2006-2008. JWPR. 2015;**5**:109-114

[9] Liu H, Zhou X, Zhao Y, Zheng D, Wang J, Wang X, et al. Factors associated with the emergence of highly pathogenic avian influenza a (H5N1) poultry outbreaks in China: Evidence from an epidemiological investigation in Ningxia, 2012. Transboundary Emerging Diseases. 2015:1-8

[10] Kammon A, Doghman M, Eldaghayes I. Surveillance of the spread of avian influenza virus type a in live bird markets in Tripoli, Libya, and determination of the associated factors. Veterinary World. 2022;**15**:1684-1690

[11] Sueur F. Stratégies d'utilisation de l'espace et des ressources trophiques par les Laridés sur le littoral picard. Rennes: Thèse Doctorat Université Rennes; 1993. p. 119

[12] Dabouineau L, Ponsero A. Synthesis on biology of common European cockle Cerastoderma edule. HAL. 2011:1-23

Chapter 2

Measuring of Cholinesterase Activity in Blood and Tissues of Bird's by Electrometric Method: Review

Ashraf S. Alias

Abstract

Many diseases and signs of organophosphate poisoning share common symptoms with other illnesses. Misusing pesticides can lead to severe damage to both the environment and people's health. Insects, birds and other animals are affected by the pesticide due to its effect on cholinesterase (ChE). This enzyme breaks down acetylcholine into smaller parts that accumulate in neuromuscular junctions and nerve terminals. In cholinergic toxicity, signs appear such as nicotine-like effects and muscarinic side effects in the nervous system. In accordance with this study, an apparatus for measuring ChE-activity in birds' blood and tissues is presented. Literature regarding proper ChE-activity in both wild and hybrid birds was found by consulting the Mosul city library. Additionally, the electrometric method proved accurate and effective. This makes it a good method for exposing potentially exposed birds in the natural world while also creating a biological control to reduce environmental pollution by carbamates or organophosphates.

Keywords: electrometric method, cholinesterase, organophosphate, pesticide misuse, poisoning

1. Introduction

Pesticides are released into the environment through a complex chain of events and can be transported through air, water, land or even within organisms affecting birds' environment. The extent of movement and spread can be local or far-reaching and varies according to the types of pesticides and their impact on the issue of food security and the provision of food for the ever-increasing world population while ignoring its effects on the birds that mainly protect crops from agricultural pests such as insects, caused their death and a significant decrease in their numbers with insect resistance to agricultural pesticides. Different types of pesticides belonging to other groups have been produced. This poses substantial risks to human health, birds and the environment [1].

Exposure of wild birds to pesticides leads to reduced egg production, which leads to a significantly decreased in the birds [2, 3]. Organophosphorus and pyrethroid compounds are used in agriculture and veterinary medicine [4, 5]. To control pest-borne diseases [6–8]. Excessive use of pyrethroids and organophosphates can cause human and animal contamination [9–13]. Organophosphate poisoning occurs due to exposure or abuse [14, 15] and intoxication occurs due to inhibition of ChE activity, leading to accumulation of acetylcholine in nerve endings and the appearance of muscarinic, nicotine and CNS symptoms-related intoxication, eventually leading to death [16, 17] but the main target of pyrethroids is the sodium channel Causes delayed blockage of the channel, resulting in a continuous flow of sodium, and its effect on voltage-sensitive chloride channels. Resulting in decreased chlorine flow through depleted airway channels [18, 19].

Measuring ChE activity in tissue and blood helps to detect organophosphate and pyrethroid toxicity in the early stages of intoxication. A 25–30% decrease in ChE activity in plasma or erythrocytes of birds is evidence of exposure such compounds [5, 9]. There appear to be several methods for measuring the activity of ChE, namely Ellman's [20, 21], radiometric [22] and electrometric [23].

The last method is simple, cheap and does not require complicated equipment (just a pH meter and a water bath). Several changes have been made to the method, such as: e.g., electrometry (Michael method), by using different buffer solutions (setup and structure), and the interaction of sample volume with different time and temperatures [24, 25]. Michael's original method was only used in humans, so some modifications were made due to various natural variations in ChE activity in the blood and tissues of different animals [24]. The main changes were to increase the incubation temperature from 20 to 37°C to increase the blood volume of the sample at the time of interaction and to minimize the incubation time depending on the animal type to which this method was applied to sheep [6, 24, 26]. Electroassay for Measuring ChE Activity in Dairy Cows [27]. The method as electrometric was not utilized previously on Budgie for measuring the activity of (ChE). Also, there is no information on the normal values of the ChE activity in birds.

2. Materials and methods

Use the electrometric method to measure the activity of (ChE) in different birds. It is important for detecting contamination with organophosphorus and carbamate pesticides that cause the death of birds by ingesting pesticides through contaminated water or food, such as fruits or insects, causing them to die. It also affects their young, causing them to die. Adequate information on normal levels of cholesterol activity in birds.

The present work uses blood and tissue samples from different birds, such as brain, liver, and muscle, ranging in sample size from 30 to 35 g. Blood samples were collected from the jugular vein (guillotine decapitation) and placed into tubes containing EDTA. Plasma was separated from blood by centrifugation at 3000 rpm for 15 minutes (Model 80-1, China). Place the brain, liver and thorax in a clean, dry plastic bag with crushed ice until homogenized at 3 mL per 100 mg of weight in barbiturate-phosphate saline (pH = 8.1) with a hand mixer. This is a sample preparation method for the purpose of screening for cholinesterase (ChE)-activity in samples that It was taken from natural birds.

3. Measure as stipulated in the method as electrometric

Blood samples and homogenized tissue were used to measure normal ChE activity. Measurements were done on the same day using electrometry **Figure 1** [27, 28], using a pH meter type (pH 025 M, US) to measure the pH 1 of the samples, and then using a water bath type (HH-2 Digital Lab Thermostatic Water Bath, China) for use. After incubation at 37°C for 30 min, samples were taken directly from the incubator and pH 2 was measured. Measure the change in acid function using the following equation:

$$(\mathrm{pH2} - \mathrm{pH1}) - \Delta\mathrm{pH\ of\ blank} = \Delta\mathrm{pH}/30\ \mathrm{min.}(\mathrm{ChE\ activity}) \tag{1}$$

ChE activity in budgie plasma and homogenate tissues was accurately measured electrometrically based on the mean and standard deviation [29], and the following coefficient of variation:

$$\mathrm{Coefficient} = \frac{\mathrm{Standard\ deviation}}{\mathrm{Mean}} \times 100 \tag{2}$$

The true and false percentages of ChE activity in the brain and plasma tissues of Budgie birds were calculated. Two sections of the plasma and brain tissue samples were separated. The first section of the samples was used to test the (ChE) activity (as previously described plus 40 µl of DW), and the second portion of the samples contained 40 µl of quinidine sulfate (0.1%) for each sample. Samples were incubated

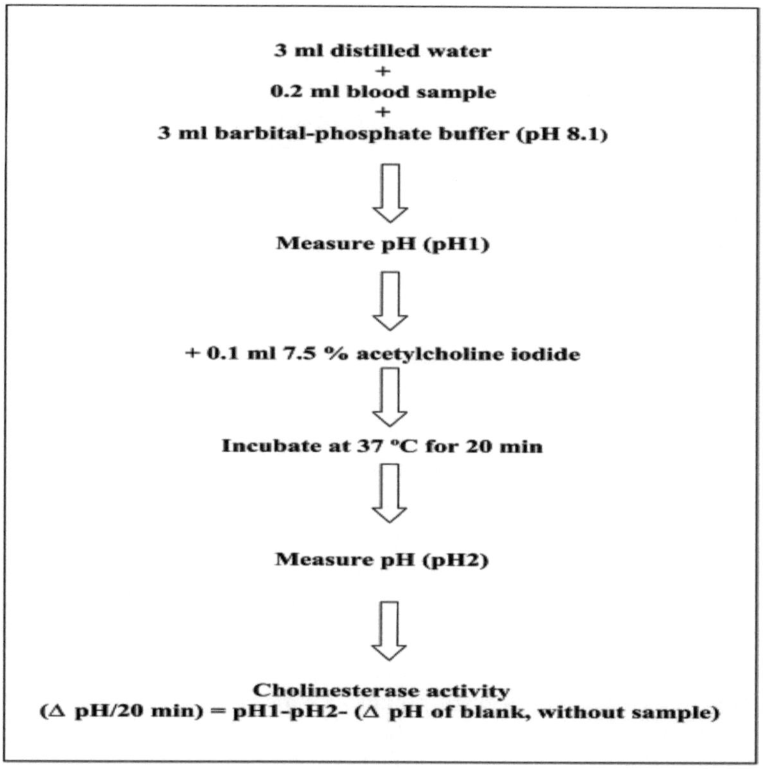

Figure 1.
Steps for the determination of blood cholinesterase activity by a modified electrometric method.

for 10 minutes at 37°C to inhibit pseudo-(ChE) [21, 30]. Through certain inhibitors, quinidine sulfate can prevent pseudo-(ChE) in brain and plasma tissues [20, 31]. Based on pseudo-(ChE) activity, the total residual (ChE) activity in brain and plasma tissues was identified as follows:

$$\begin{array}{c}(\text{ChE})\ \text{activity} = \text{total}\,(\text{ChE}) - \text{activity of real}\,(\text{ChE}) \\ (\text{With no quinidine})\,(\text{following quinidine addition})\end{array} \tag{3}$$

Dichlorvos 50% (produced in China by Shenzhen Horizon Industry Co. Ltd.) and deltamethrin 2.5% were used to assess the inhibition of activity (ChE) in the brain tissues of Budgie birds in vitro (Indogulf company, India).

After obtaining the brains of seven Budgie birds for each experiment and incubating the samples for 10 minutes at 37°C to inhibit activity, inhibitor (ChE) incubation was carried out with concentrations of 0 (control), 0.5, and 1 μM/L for the treated group of Dichlorvos and concentrations of 0, 5 and 10 μM/L for the group of Deltamethrin [32]. Before inhibition, the percentage was calculated as follows:

$$\begin{array}{l}\text{Inhibition \% (ChE) activity} = \\ \dfrac{\begin{array}{l}\text{Activity of (ChE) for control sample (with no dichlorovos or deltamethrin)} - \\ \text{Activity of (ChE) with Dichlorovos or deltamethrin}\end{array}}{\text{Activity of (ChE) for control sample (with no dichlorovos or deltamethrin)}} \times 100\end{array} \tag{4}$$

4. Results

Table 1 shows the normal reference range values, 95% confidence interval and related statistics for plasma, brain, liver and pectoralis muscle ChE activities of the bird species studied. Using the technique of in vitro ChE inhibition for 10 min by 0.1% quinidine sulfate, the estimated percentages of true ChE activity in the plasma of the Quail, Large pin-tailed sand grouse, Starling, Rock dove (*Columba livia*), Local dove (*Streptopelia Senegalensis*) and Budgie birds were 77%, 69%, 71% and 73%, 75%, 75%, respectively (**Table 2**).

5. Conclusion

We conclude from the foregoing that measuring cholinesterase activity in the blood (plasma) and tissues (brain, liver, and muscles) of wild birds and domestic birds that live with humans is very important through which we discover the values of cholinesterase activity in wild birds, which reveal to us the extent of environmental pollution with these pollutants. (Pesticides) and their effect on wildlife, especially endemic and migratory birds, which cause the sudden death of large numbers of birds when these birds migrate through countries that use pesticides randomly and unstudied. Protecting our environment from pollution with organic phosphorus pesticides and carbamate pesticides that cause the death of birds directly or indirectly by poisoning and killing the organisms that birds depend on for their food, which causes the death of birds and the loss of this great environmental wealth. Because it maintains the ecological balance and is also a natural agricultural pesticide.

Individuals per species					
	Cholinesterase activity (ΔpH/30 min)				
	Plasma	**Brain**	**Liver**	**Muscle**	**Reference**
Quail					
Mean	1.23	0.39	0.19	0.06	[33]
Standard Error	0.098	0.013	0.021	0.004	
Standard Deviation	0.310	0.040	0.066	0.013	
95% confidence Interval	1.008, 1.452	0.360, 0.418	0.124, 0.146	0.050, 0.068	
Range	1.71–0.72 = 0.99	0.45–0.33 = 0.12	0.34–0.09 = 0.25	0.08–0.04 = 0.04	
Sand grouse					
Mean	1.81	0.37	0.06	0.07	[33]
Standard error	0.109	0.025	0.009	0.005	
Standard deviation	0.345	0.081	0.029	0.016	
95% confidence interval	1.512, 2.068	0.304, 0.434	0.037, 0.068	0.054, 0.077	
Range	2.58–1.44 = 1.14	0.51–0.25 = 0.26	0.12–0.01 = 0.11	0.09–0.04 = 0.05	
Starling					
Mean	1.10	0.24	0.08	0.08	[33]
Standard Error	0.074	0.029	0.006	0.008	
Standard deviation	0.235	0.093	0.019	0.025	
95% confidence Interval	0.936, 1.272	0.174, 0.308	0.066, 0.081	0.057, 0.093	
Range	1.56–0.84 = 0.72	0.38–0.12 = 0.26	0.11–0.04 = 0.07	0.13–0.04 = 0.09	
Rock dove (*Columba livia*)					
Mean	1.28	0.59	0.12	0.08	[34]
Standard Error	0.08	0.026	0.010	0.008	
Standard deviation	0.256	0.084	0.030	0.025	
95% confidence Interval	1.100, 1.466	0.525, 0.585	0.094, 0.138	0.062, 0.099	
Range	1.70–0.85 = 0.85	0.75–0.45 = 0.28	0.17–0.07 = 0.10	0.14–0.05 = 0.09	
Local dove (*Streptopelia Senegalensis*)					
Mean	0.82	0.39	—	—	[35]
Standard error	0.012	0.014	—	—	
Standard deviation	0.038	0.044	—	—	
95% confidence Interval	0.527, 0.587	0.302, 0.432	—	—	
Range	1.68–0.90 = 0.78	0.65–.28 = 0.37	—	—	

Individuals per species					
	Cholinesterase activity (ΔpH/30 min)				
	Plasma	**Brain**	**Liver**	**Muscle**	**Reference**
Budgie birds					
Mean	0.31	0.23	0.12	0.11	[36]
Standard error	0.005	0.003	0.003	0.002	
Standard deviation	0.015	0.009	0.007	0.005	
95% confidence interval	0.30–0.32	0.22–0.23	0.11–0.12	0.10–0.11	
Range	0.34–0.30 = 0.04	0.24–0.21 = 0.03	0.13–0.11 = 0.02	0.11–0.10 = 0.01	

Cholinesterase values are the mean ± SE of 10 plasma samples for each of bird species.

Table 1.
*The cholinesterase activity values listed in this table were found out by an electrometric method in the bird species Quail, sand grouse, Starling, Rock dove (*Columba livia*), Local dove (*Streptopelia Senegalensis*) and Budgie birds. N = 10.*

Bird	**Variable**	**(ΔpH/30 minutes)**	**Activity (%)**	**Reference**
Quail	Total cholinesterase	1.230 ± 0.093	100	[33]
	True cholinesterase*	0.950 ± 0.069	77	
	Pseudo cholinesterase	0.280 ± 0.054	23	
Large pin-tailed sand grouse	Total cholinesterase	1.810 ± 0.109	100	[33]
	True cholinesterase*	1.240 ± 0.076	69	
	Pseudo cholinesterase	0.570 ± 0.126	31	
Starling	Total cholinesterase	1.100 ± 0.074	100	[33]
	True cholinesterase*	0.780 ± 0.025	71	
	Pseudo cholinesterase	0.320 ± 0.042	29	
Rock dove (*Columba livia*)	Total cholinesterase	1.280 ± 0.080	100	[33]
	True cholinesterase*	0.930 ± 0.063	73	
	Pseudo cholinesterase	0.350 ± 0.028	27	
Local dove (*Streptopelia Senegalensis*)	Total cholinesterase	0.03 ± 0.83	100	[35]
	True cholinesterase*	0.03 ± 0.62	75	
	*Pseudo cholinesterase	0.018 ± 0.21	25	
Budgie birds	Total cholinesterase	0.31 ± 0.005	100	[36]
	True cholinesterase*	0.002 ± 0.255	75	
	Pseudo cholinesterase	0.06 ± 0.006	25	

Cholinesterase values are the mean ± SE of 10 plasma samples for each of bird species.
**Quinidine sulfate was used to inhibit pseudo cholinesterase activity in the plasma.*

Table 2.
*Estimation of true cholinesterase activity as determined by the described electrometric method in the plasma of Quail, Sand grouse, Starling, Rock dove (*Columba livia*), Local dove (*Streptopelia Senegalensis*) and Budgie birds.*

Acknowledgements

For their assistance and completion of this work, the author is grateful to the Environmental Science Department of the College of Environmental Science and Technologies at Mosul University in Iraq.

Author details

Ashraf S. Alias
Environmental Science Department, Environmental Science and Technologies Mosul University, United States

*Address all correspondence to: ashraf.saddik.@uomosul.edu.iq

References

[1] Karen DJ, Joab BM, Wallin JM, Johnson KA. Partitioning of chlorpyrifos between water and an aquatic macrophyte (Elodea densa). Chemosphere. 1998;**37**(8):1579-1586. DOI: 10.1016/S0045-6535(98)00141-6

[2] Weir SM, Youssif MR, Anderson T, Salice CJ. Current progress in developing standardized methods for reptilian toxicity testing to inform ecological risk assessment. Bird and Reptile Species in Environmental Risk Assessment Strategies. 2023:130-150

[3] Singh PP, Kumar A, Chauhan RS, Pankaj PK. How safe is the use of chlorpyrifos: Revelations through its effect on layer birds. Veterinary World. 2016;**9**(7):753-758. DOI: 10.14202/vetworld.2016

[4] Bradberry SM, Cage SA, Proudfoot AT, Veale JA. Poisioning due to pyrethroid. Toxicological Review. 2005;**24**(2):93-106. DOI: 10.2165/00139709-200524020-00003

[5] Thompson CM, Prins JM, George KM. Mass spectrometric analyses of organophosphate insecticide oxon proteinadducts. Environmental Health Perspectives. 2009;**118**(1):11-19. DOI: 10.1289/ehp.0900824

[6] Alias A, Megdad MH, Ali MH, Esmaial YA. Study of (ChE)activity in adult ewes treated by the anthelmintic drug levamisole. Al-Rafidain Science Magazine. 2010;**21**(3):93-102. Available from: https://scholar.google.com/citations?

[7] Kazal EH, Alias AS. Survey of some therapeutic formulation for anti-ectoparasites used in clinic veterinary in Tilkaif and clinic veterinary teaching Hospital in Nineveh province. Al-Rafidain. Science Magazine. 2014;**25**(1):57-66. Available from: https://www.iasj.net/iasj?func=article&aId=86108

[8] Tomlin CD. The Pesticide Manual: A World Compendium. 14th ed. Alton, Hampshire, UK: British Crop Protection Council; 2006. pp. 186-187. Available from: https://www.scirp.org/(S(lz5mqp453edsnp55rrgjct55))/reference/ReferencesPapers.aspx?ReferenceID=1880037

[9] Kumar S, Kaushik G, Villarreal-Chiu JF. Scenario of organophosphate pollution and toxicity in India: A review. Environmental Science and Pollution Research. 2016;**23**:9480-9491. DOI: 10.1007/s11356-016-6294-0

[10] Vermeire T, Mcphail R and Waters M. Organophosphorous pesticides in the environment. In World Health Organization. IV. Meeting Report of the International Workshop on Approaches to Integrated Risk Assessment 2001; pp. 1-18. Available from: https://www.who.int/ipcs/publications /en/ch_3d.pdf

[11] Lotti M. (ChE) inhibition : Complexities in interpretation. Clinical Chemistry. 1995;**41**(12 Pt 2):1814-1818. Available from: https://www.ncbi.nlm.nih.gov/pubmed/7497638

[12] Ahmed A, Kai I, Wang JR, Olson MR, Bonner OH, Gaafar AR, et al. The impact of repeated organophosphorus pesticide exposure on biomarkers and neurobehavioral outcomes among adolescent pesticide applicators. Journal of Toxicology and Environmental Health, Part A. 2017;**80**(10-12):542-555. DOI: 10.1080/15287394. 2017. 1362612

[13] Haug G, Hoffiman H. Chemistry of plant protection. In: Synthetic

Pyrethroid Insecticides: Structures and Properties. New York: Springer-Verlag; 1990

[14] Alias AS. The use of largactil as an anti-poisoning with organopgosphorus pesticides. Veterinary Practitioner. 2021; **22**(2):137-139

[15] Ahmed AS. Change in acetylcholine activity and some blood parameters in adult sheep dipped in deltamethrin. Iraqi Journal of Veterinary Sciences. 2020;**35** (2):301-304

[16] Moretto A. Experimental and clinical toxicology of anti (ChE). Toxicology Letters. 1988;**28**(102–103):509-513. Available fom: https://www.ncbi.nlm.nih.gov/pubmed/10022304

[17] Tilson HA. New horizons: Future directions in neurotoxicology. Environ. Hlth Perspect. 2000;**108**:439-441. DOI: 10.1289/ehp.00108s3439

[18] Sderlund DM, Clark JM, Sheets LP, Mullin LS, Piccirillo VJ, Sargent D, et al. Mechanisms of pyrthroid neurotoxicity: Implications for cumulative risk assessment. Toxicology. 2002;**171**(1):3-59. DOI: 10.1016/s0300-483x(01) 00569-8

[19] Dong K. Insect sodium chaaels and insecticides resistance. Invertebrate Neuroscience. 2007;**7**(1):17-30. DOI: 10.1007/s10158-006-0036-9

[20] Ellman GL, Courtney KD, Andres V, Featherstone RM. A new and rapid colorimetric determination of acetyl (ChE) activity. Biochemical Pharmacology. 1961;7:88-95. DOI: 10.1016/0006-2952(61)90145-9

[21] Hestrin S. The reaction of acetylcholine and other carboxylic acid derivatives with hydroxylamine, and its analytical application. Journal of Biological Chemistry. 1949;**180**(1):249-261. Available from: https://www.ncbi.nlm.nih.gov/pubmed/18133390

[22] Johnson CD, Russel RL. A rapid simple radiometric assay for (ChE) suitable for multiple determinations. Analytical Biochemistry. 1975;**64**:229-238. DOI: 10.1016/j.taap.2009.02.022

[23] Michel HO. An electrometric method for the determination of red blood cell and plasma cholinesterase activity. Journal of Laboratory and Clinical Medicine. 1949;**34**:1564-1568. DOI: 10.1007/BF03161035

[24] Mohammad FK, St Omer VE. Modifications of Michel's electrometric method for rapid measurement of blood cholinesterase activity in animals: A minireview. Veterinary and Human Toxicology. 1982;**24**(2):119-121

[25] Silvestri R. New techniques to measure blood cholinesterase activity in domesticated animals. American Journal of Veterinary Research. 1977;**38**(5):659-662. Available from: https://www.ncbi.nlm.nih.gov/pubmed/18075

[26] Mohammad FK, Faris GA, Al-Kassim NA. A modified electrometric method for measurement of erythrocyte acetylcholinesterase activity in sheep. Veterinary and Human Toxicology. 1997;**39**(6):337-339. Available from: http://europepmc.org/abstract/med/9397501

[27] Mohammad FK, Faris GAM, Alias AS, Al-Baggou B. Blood cholinesterase activities in cattle, sheep and goats measured by a modified electrometric method. Journal of Animal and Veterinary Advances. 2005;**4**:923-926. Available from: http://docsdrive.com/pdfs/medwelljournals/javaa/2005/923-926.pdf; https://www.Researchgate.net/publication/26591158_Blood_(ChE)_

Activities_n_Cattle_Sheep_andGoats_Measured_by_a_Modified_Electrometric_Method

[28] Mohammad FK, Alias AS, Ahmed OAH. Electrometric measurement of plasma, erythrocyte, and whole blood cholinesterase activities in healthy human volunteers. Journal of Medical Toxicology. 2007;**3**:25-30

[29] Scarsella G, Toschi G, Bareggi SR, Giacobini E. Molecular forms of cholinesterases in cerebrospinal fluid, blood plasma, and brain tissue of the beagle dog. Journal of Neuroscience Research. 1979;**4**(1):19-24

[30] Jabouri A, Hassan MM. Determination of (ChE) Activity in Goat Blood by Method as Electrometrics: Inhibition of Organophosphorus and Carbamate Pesticides. Mosul, Iraq: University of Mosul; 2004. DOI: 10.3390/ s18124281

[31] Abass K. Validation of an Eiectrometeric method for cholinesterase measurement in the plasma of ducks. Tikrit Journal of Pure Science. 2006;**11**(1):1-4. Available from: https://www.iasj.net/iasj?func=article&aId=39819

[32] Faris GA-M, Al-Dewachi OS, Said MO, Mohammad FK. Determination of plasma cholinesterase activity in cockerels by an electrometric method. Iraqi Journal of Veterinary Sciences. 1999;**12**:255-260

[33] Alias AS, Mohammad FK. Electrometric measurement of plasma and tissue cholinesterase activities of four wild birds in Iraq. Journal of Medicl Toxicology. 2005;**4**:197-202. DOI: 10.2478/v10102-011-0022-x

[34] Alias S. Effect of dichlorvos on cholinesterase activity in pigeons (rock dov). Iraqi Journal of Veterinary Sciences. 2006;**20**(2):191-202

[35] Alias AS. The use of an electrometric method for measurement of cholinesterase activity in plasma and tissues of local doves. Iraqi Journal of Veterinary Sciences. 2009;**23**(1):9-14

[36] Alias AS. Measuring of cholinesterase activity via electrometric method in blood and tissues of budgie bird's. Veterinary Practitioner. 2020;**21** (2):P450-P454

Chapter 3

Anthropogenic Noise Affects the Bird Song Frequency and Behavioral Response: A Review

Garima Singh, Sachin Kumar, Kalpana Chaudhary and Gaurav Sharma

Abstract

Nowadays, as anthropological disturbance increases in animal's life, they are adjusting their nature to a novel environment. Birds have also severe constrain on vocal communication by interfering with selection of acoustics signals. Noise from urban area affects negatively bird fitness with their communication and, for instance, activity related to breeding also had bad impact on predator detection. It influences also local distribution patterns as well as bird communities due to continuing exposure. Sometimes birds can survive with urban noise, but mechanism remains unknown. Here, we focused on relation between firstly anthropogenic noise and bird richness secondly noise level and song modification and then species noise tolerance and detection frequency. This kind of noise may hamper recognition of song by female, makes difficulties in territory marking, and also affects the maintenance of pair bond in birds. Studies on the quantification and documentation of acoustical characteristics and structural variability in birdsong provide substantial information on its conceptual and empirical significance. Songs may vary at local level in neighboring groups of birds and the level of variations depends on selection, based on various behavioral and ecological factors. So, it is important to understand the vocal communication for successful breeding for the conservation and for maintaining a balance within ecosystem.

Keywords: avian acoustics, urban climate, behavioral response, anthropogenic noise, sound frequency

1. Introduction

Avian habitat is becoming increasingly scarce due to rising populations. They have a deleterious impact on communication, territorial behavior, foraging, and reproduction in the entire animal community. The vocal cord of songbirds may be affected by anthropogenic interference as urbanization increases in cities. Higher-frequency song of blackbird (*Turdus merula)* components had greater amplitudes as well, and city birds tended to sing higher-frequency (and consequently higher amplitude) song elements over forest species [1]. Anthropogenic noise is a pervasive pollutant that changes how animals communicate acoustically. Some animals adjust their vocalizations in response to environmental noise. There is still much to discover about how much adjustment efficacy varies depending on the source of the noise, if signal adjustments improve communication in noise, and how individual physiology variance varies depending on reaction time. After assessing each subject's adrenocortical response, songs that had been modified for noise and those that had not been modified were played to wild Savannah Sparrows (*Passerculus sandwichensis*). Under some altered acoustic conditions, the playback of songs that were rewritten for loud surroundings was able to successfully reestablish suitable conspecific territorial hostile behavior [2].

Finding a link between adrenocortical responsiveness levels that reduced communication mistakes in some types of infrastructure and increased errors in others was unanticipated. Song alterations that worked well at pump jacks did not function well at screw pumps for persons with poorer adrenocortical reactivity and vice versa. These results demonstrate that vocal adaptations can sometimes compensate for communication gaps caused by environmental noise, but physiological variations among receivers may limit the use of these adaptations. Mitigation strategies must consider both the acoustic and physiological effects of infrastructure in order to reduce anthropogenic noise.

Acoustic communication can be disrupted by noise, which can then have an impact on signaling behavior and the development of acoustic communication [3]. The Lombard effect, cultural transmission, an unintentional change in vocal effort, or a decisive action by the signaler are examples of mechanisms of selection that can result in changes in acoustic behavior as a result of efforts to enhance signal detection and discrimination in the presence of background noise (e.g. adjustment in timing of vocalization) [4–6]. Immediate signaling flexibility (ISF), a context-dependent type of behavioral plasticity, enables animals to change their acoustic behavior momentarily in response to changes in background noise. This can involve adjusting the amplitude, spectral or temporal features of acoustic signals, or the timing of signaling behavior [7, 8]. However, the way and degree to which the signal is altered vary between and within species depending on the species' capacity for behavioral flexibility, developmental plasticity, and micro-evolutionary responses as well as the flexibility of the signaler, their own flexibility, their past experiences, and their perception of noise [9].

2. Amplitude increases with frequency

Nemeth found that in the frequency range of up to 2.2 kHz, blackbird song components clearly showed a positive association between frequency and amplitude. The amplitude varied significantly with frequency; for instance, the average maximum amplitude level increased by more than 15 dB from 1.5 to 2.5 kHz. Blackbird song showed a coupling between peak frequency and amplitude, and city birds tended to sing higher-frequency components that can be produced at louder sound intensities. Both the raised frequency and the corresponding increase in amplitude lessen the acoustic masking caused by low-frequency traffic noise, but the frequency-dependent amplitude change has a stronger impact. City birds may be able to sing at larger amplitudes to lessen acoustic masking by noise by selecting higher elements [10]. This discovery supports earlier research that found positive relationships between frequency and amplitude in other songbird species [11]. This finding implies that nearby mechanisms, such as physical impedances [12], biophysical constraints [13], or physiological restraints [14], maybe what prevent the production of loud vocalizations at the lower end of the frequency range.

Noise's detrimental effects on reproduction may be linked to its effects on communication. Anthropogenic noise may be interfering with mating selection due to masking effects and the need to modify communication [15]. Additionally, we are aware that parental and child interaction is crucial for the survival of the latter [16]. For instance, begging is a behavior used by nesting passerine birds to signal their need for food. Begging involves a combination of postures and sounds.

Studies have indicated that anthropogenic airborne noise affects auditory communication [17]. Synthetic traffic stimuli (spectral peak: 10 Hz) elicited a call-rate change at a lower threshold than synthetic wind-turbine playbacks, whereas the former did not (spectral peak: 100 Hz).

3. Discussion

It is hypothesized that noise sources that are both novel or unpredictable and related to a physiologically significant sound will elicit reactions resembling those linked to predation risk [18]. Therefore, cutting back on or stopping calling could increase the likelihood of avoiding the predator's attention. Furthermore, compared to a multimodal stimulus, a particular univocal stimulus supplied to the animals may be ineffective [19]. No matter the reason, a drop in calling rate can have repercussions. Numerous scientists have claimed that elevated airborne noise levels have negative effects on animal communication during mating, which can affect reproductive success [20]. It also has a negative impact on species interaction within population and male–female response and playback song behavior.

According to projections, an additional two billion people will require a place to live in the next 20 years, but they will not reside in the existing cities but rather in newly built urban regions [21]. Rare species are typically negatively impacted by urbanization, while more widespread species that can include native generalist species but frequently refers to non-native urban colonizers are favored [22–25]. Diversified urban ecology does offer a wide range of niches, therefore urbanization does not always result in a loss of species diversity [26]. However, on a broader scale, urbanization results in homogenization and a loss of diversity since, regardless of the original fauna, typical urban species end up being the same everywhere [27, 28]. Urban problems may be particularly stressful for animals that still exist in cities and outside of them, as evidenced, for example, by a divergence in heterophil-leucocyte ratios [29] or elevated baseline corticosterone levels in male birds in urban habitats [30].

However, despite the fact that many species leave cities because they depend on habitat elements that no longer exist, others find a new home among the bricks and concrete and adjust to life in the city [31]. Many animal groups utilize acoustic signaling to protect their territories, alert conspecifics to oncoming predators, or attract mates [32]. Individual fitness and population persistence may suffer from any changes to the transmission environment that prevent acoustic signals from reaching the intended receiver or alter the information content of the signal. Chronic noise exposure is a result of expanding transportation networks, resource exploitation, motorized recreation, and urban expansion, even in the most remote wilderness locations [33, 34]. Only if signalers change the form of their songs or if receivers are able to shift their perceptual apparatuses in reaction to background noise can effective animal communication be maintained in altered auditory settings [9]. When background noise and a particular species' auditory output coincide in their frequency spectra at a specific time and location, acoustic masking may result.

Animals frequently employ acoustic communication to share information among themselves. It is done by using an auditory signal that is produced by a sender, travels through the environment, and is picked up by a receptor [35]. Acoustic waves can travel great distances and carry data, including the sender's identity, position, and sexual orientation. Their use is appropriate in a variety of situations, such as luring mates for breeding, defending one's territory, and warning of danger. Interactions between communicators are hampered when the communication process is inefficient from signal emission to signal reception. Thus avian species in urban climates have a high frequency in their song in comparison to non-urban climates. They have to produce high pitch sounds for a song, affect male–female interaction and behavioral responses due to anthropogenic noise, and somehow indirectly affect the breeding.

4. Conclusion

Anthropogenic noise is a category of sound that can interfere with communication and be viewed as a form of environmental pollution. The next critical step in determining the true ecological effects of noise pollution is more research into how species in the wild not only detect but also distinguish between signals within noise gradients. Therefore, it is crucial to comprehend how noise pollution might contribute to further habitat degradation for sensitive species already suffering from habitat loss and climate change, as well as how ecological effects from acoustic environment modification compare to anthropogenic effects that have drawn ecologists' attention for a longer period of time. In the future, we propose developing studies to disentangle the distinctions between distraction and masking, to investigate how an organism's behavior may be influenced by anthropogenic noise that is outside of its vocal range. To examine how a greater variety of song types affect issues of detection and discrimination and to better understand demographic effects, we also recommend including examples with vocal qualities that deviate from normative conditions. According to playback trials, anthropogenic vibratory stimuli significantly decreased the calling activity in focused males, with their mean call rate falling by 50%. Vibrations from wind turbines and vehicles had an equivalent effect on call volume. The results of comparing the prerecorded and artificial stimuli show that the whole vibrational spectrum, not just its peak frequency, was responsible for the observed responses. Additionally, playback of naturally recorded audio decreased the sound strength threshold at which animals changed their baseline calling behavior and other interaction in populations toward each other. Future research, particularly field-based experiments in appropriate systems, will allow for a better understanding of the full extent of anthropogenic noise's effects on species and communities, how the acoustic environment has changed in comparison to other well-studied human-induced habitat modifications, and how these effects might be mitigated through avian conservation measures that could be viewed as a form of environmental pollution and that could potentially help.

Author details

Garima Singh[1*], Sachin Kumar[2], Kalpana Chaudhary[3] and Gaurav Sharma[1]

1 Zoological Survey of India, Dehradun, Uttrakhand, India

2 Punjab Agriculture University, Ludhiana, Punjab, India

3 R.G. (P.G.) College Meerut, UP, India

*Address all correspondence to: garimasinghzoology@gmail.com

References

[1] Riede T, Suthers RA, Fletcher NH, Blevins WE. Songbirds tune their vocal tract to the fundamental frequency of their song. Proceedings of the National Academy Science USA. 2006;**103**:5543-5548

[2] Marzluff JM. Island biogeography for an urbanizing world: How extinction and colonization may determine biological diversity in human-dominated landscapes. Urban Ecosystem. 2005;**8**:157-177

[3] Patricelli GL, Blickley JL. Avian communication in urban noise: Causes and consequences of vocal adjustment. The Auk. 2006;**123**:639-649

[4] Rabin LA, Greene CM. Changes to acoustic communication systems in human-altered environments. Journal of Comparative Physiology. 2002;**116**:137-141

[5] Ryan MJ, Kime NM. Selection on long distance acoustic signals. In: Simmons AM, Fay RR, Popper AN, editors. Acoustic Communication. Berlin: Springer-Verlag; 2003. pp. 225-274

[6] Dabelsteen T. An analysis of the full song of the blackbird *T. merula* with respect to message coding and adaptations for acoustic communication. Ornis Scandinavica. 1984;**15**:227-239

[7] Bolger DT, Scott TA, Rotenberry JT. Breeding bird abundance in an urbanizing landscape in coastal southern California. Conservation Biology. 1997;**11**:406-421

[8] Narins PM, Grabul DS, Soma KK, Gaucher P, Hödl W. Cross-modal integration in a dart-poison frog. Proceedings of the National Academy Science USA. 2005;**102**:2425-2429

[9] Francis C, Barber JR. A framework for understanding noise impacts on wildlife: An urgent conservation priority. Frontiers in Ecology and Evolution. 2013;**11**:305-313

[10] Hans S, Ripmeester EAP. Birdsong and anthropogenic noise: Implications and applications for conservation. Molecular Ecology. 2008;**17**:72-83

[11] Curry CM, Paulson G, Des B, Patricia NK. Noise source and individual physiology mediate effectiveness of bird songs adjusted to anthropogenic noise. Scientific Reports. 2018;**8**:3942

[12] Shannon G. A synthesis of two decades of research documenting the effects of noise on wildlife. Biological Reviews. 2016;**91**:982-1005

[13] Vitousek PM, Mooney HA, Lubchenco J, Melillo J. Human domination of earth's ecosystems. Science. 1997;**277**:494-499

[14] Narins PM, Meenderink SWF, Tumulty JP, Cobo-Cuan A, Márquez R. Plant-borne vibrations modulate calling behavior in a tropical amphibian. Current Biology. 2018;**28**:R1333-R1334

[15] Harding HR, Gordon TAC, Eastcott E, Simpson SD, Radford AN. Causes and consequences of intraspecific variation in animal responses to anthropogenic noise. Behavioral Ecology. 2019;**30**:1501-1511

[16] Nemeth E, Pieretti N, Zollinger SA, Geberzahn N, Partecke J, Mirand AC, et al. Bird song and anthropogenic noise: Vocal constraints may explain why birds

sing higher-frequency songs in cities. Proceedings of the Royal Society B. 2013;**280**:20122798

[17] Senzaki M, Nakamura F, Kadoya T, Francis CD, Ishiyama N. Suffering in receivers negative effects of noise persist regardless of experience in female anurans. Functional Ecology. 2018;**32**:2054-2064

[18] Mohneke R, Schneider H. Effect of temperature upon auditory thresholds in two anuran species, Bombina v. variegata and Alyteso. obstetricans (Amphibia, Discoglossidae). Journal of Comparative Physiology. 1979;**130**:9-16

[19] Ríos-Chelén AA. Bird Song: The Interplay between Urban Noise and Sexual Selection. Acta Oecologica. 2009;**13**(01):153-164

[20] McKinney ML. Urbanization as a major cause of biotic homogenization. Biological Conservation. 2006;**127**:247-260

[21] Meyer WB, Turner B. Human population growth and global land use/cover change. The Journal of Animal Ecology. 1992;**23**:39-61

[22] Slabbekoorn H. Songs of the city: Noise-dependent spectral plasticity in the acoustic phenotype of urban birds. Animal Behaviour. 2013;**85**:1089-1099

[23] Penna M, Velásquez NA, Bosch J. Dissimilarities in auditory tuning in midwife toads of the genus Alytes (Amphibia: Anura). Biological Journal of the Linnean Society. 2015;**116**:41-51

[24] Bosch J, Márquez R. Female preference intensities on different call characteristics and symmetry of preference above and below the mean in the Iberian midwife toad Alytes cisternasii. Ethology. 2015;**111**:323-333

[25] Luther D, Gentry K. Sources of background noise and their influence on vertebrate acoustic communication. Behavior. 2013;**150**:1045-1068

[26] Clergeau P, Croci S, Jokimäki J. Avifauna homogenization by urbanization: Analysis at different European latitudes. Biological Conservation. 2006;**127**:336-344

[27] Ruiz G, Rosenmann M, Novoa FF, Sabat P. Hematological parameters and stress index in rufous-collared sparrows dwelling in urban environments. Condor. 2002;**104**:162-166

[28] Gerhardt HC, Bee MA. Recognition and localization of acoustic signals. In: Narins PM, Feng AS, Fay RR, Popper AN, editors. Hearing and Sound Communication in Amphibians. Springer; 2007. pp. 113-146

[29] Barber JR, Crooks KR, Fristrup KM. The costs of chronic noise exposure for terrestrial organisms. Trends in Ecology & Evolution. 2010;**25**:180-189

[30] Sewell SR, Catterall CP. Bushland modifications and styles of urban development: Their effects on birds in southern Queensland. Wildlife Research. 1998;**25**:41-63

[31] Kleist NJ, Cruz GA, Francis CD. Anthropogenic noise weakens territorial response to intruder's songs. Ecosphere. 2016;**7**(3):e01259

[32] Mason MJ, Narins PM. Vibrometric studies of the middle ear of the bullfrog (Rana catesbeiana) II. The operculum. Journal of Experimental Biology. 2002;**205**:3167-3176

[33] Brumm H, Slabbekoorn H. Acoustic communication in noise. Advances in the Study of Behaviour. 2005;**35**:151-209

[34] Brumm H, Zollinger SA. The evolution of the Lombard effect: 100 years of psychoacoustic research. Behavior. 2011;**148**:1173-1198

[35] Bonier F, Martin PR, Sheldon KS. Sex-specific consequences of life in the city. Behavioral Ecology. 2007;**18**:121-129

Chapter 4

An Approach to the Natural and Induced Evolution of Wetlands at La Mancha Húmeda (Central Spain) – We Are Arriving Too Late

Luis Fernando Basanta Reyes

Abstract

An important part of the migratory population of european waterfowl, uses Spanish wetlands during their migratory trips and/or winter periods. In Spain, resident and breeding birds of this group, are important due to its number and some endangered species. Current populations have suffered a great decline since the middle of the last century, attributed, among other reasons, to the loss in number and quality of national wetlands. An area located in the center of Spain has been selected: a large plain called "La Mancha". The number of lagoons, different ecological types and the "tablas" or river overflows, have constituted the values to be recognized as one of the most important Ramsar sites in Europe, and a Biosphere Reserve. Evolution have been studied, due to natural causes and human actions: trying to eradicate malaria and "recovering" land for agriculture, draining it through canals, ditches and wells, using it as landfill for solid waste and fecal and industrial water, and the extraction of water from undergroundb aquifers. These actions have destroyed, forever, 80% of the wetlands in this area. The effects on the associated fauna have been direct: its disappearance. The author is a direct witness to the development state of these wetlands over the last 50 years (1972-2023).

Keywords: wetlands, Mancha Húmeda, Castilla-La Mancha, waterfowl, drainage, aquifer

1. Introduction

Talking about wetlands in Spain might seem incongruous to the inhabitants of neighboring European countries and even to the majority of Spaniards themselves. The image of a country with many sunny days and relatively few rainy days does not invite to imagine large flocks of ducks or waders crossing the sky, feeding on the mud of shores or diving in a lagoon. That's how it was and is, but the description of "large flocks" belongs to the past, sadly.

Since ancient times, man has considered wetlands as gloomy places and emanating miasmas, "putrid odor" and "bad vapors." Spain has not been an exception, with abundant examples.

At the beginning of the nineteenth century (and several centuries before, but without sufficient means), civil engineers began to plan and execute sanitation works that especially affected those wetlands located in the vicinity of villages. The direct relationship of malaria with thousands of human deaths over the years cannot be denied, mainly in settlements located in the vicinity of flooded areas, some of these populations having to be completely abandoned. What makes these drying measures understandable [1, 2].

Malaria was considered endemic in Spain until 1964. Improvements in the knowledge of the disease, hygienic and sanitary conditions of the population, and medication, led the WHO to declare malaria eradicated throughout the national territory [3, 4].

Decisions to drain lagoons and channel rivers were not only motivated by the responsibility of governments, interested in alleviating diseases in the population; they also lobbied farmers' interests, claiming crop losses due to flooding in flood cycles. In addition, there was interest in occupying the dried-up land that promised to be very fertile. It was not in most cases, since they turned out to be areas with high salinity, making the emerged lands unproductive for cultivation. On the other hand, if the drying up was sponsored by an individual, the State grant him the exploitation of the dried-up land for 99 years [5–8].

According to data from the Ramsar Convention Bureau, since 1700 almost 90% of the world's wetlands have been lost. During the twentieth century alone, the global extent of wetlands decreased by 64–71%. Between 1970 and 2015 the reduction was 35% (three times higher than the loss of forests) [9].

In Spain, most of the natural wetlands that disappeared until the 1960s and 1970s were drained and transformed, fundamentally, for agricultural use. It is estimated that, at that time, between 60% and 70% of the original surface had already disappeared. Between the years 1948 and 1990, a loss of 60% is calculated. These figures are similar to those from other countries in our socioeconomic style [10, 11].

Wetlands have been continually threatened and destroyed in Spain, throughout Europe... throughout the world.

This chapter reveals some of the wetlands that existed and were lost for different reasons. It does not forget the treatment that the wetlands of Castilla-La Mancha have received in the past and even in recent times. A summarized follow-up of the evolution of the existing ones is carried out. At the same time, it suggests recommendations considered important and basic, when recovering wetlands or creating new artificial lagoons.

The intention is not to raise wounds, some bad or never healed. It is intended, without harshness, to relate and analyze some of the mistakes that were made at certain moments in the past, trying to amend them in the present and avoid them in the future.

It is fair to recognize that, at present, interest in the conservation of wetlands is infinitely greater on the part of the general public, researchers and officials of the Administration of Castilla-La Mancha and Spain, with very notable compromise examples (**Figures 1** and **2**).

Figure 1.
Winter flock of coots (Fulica atra). Photo A. Amor.

Figure 2.
Winter flock of red-crested pochard (Netta rufina) and coots (Fulica atra). Photo L. Basanta.

2. The geographical space of Castilla-La Mancha

2.1 La Mancha húmeda

Castilla-La Mancha is a region located in the center of Spain, politically called the Autonomous Community. It is made up of five provinces: Albacete, Ciudad Real, Cuenca, Guadalajara and Toledo. It is the third largest Spanish region. It occupies an area of 79,409 km^2, with well-differentiated plains and mountains. It represents 15.7% of the Spanish territory [12, 13].

Wetlands of Castilla-La Mancha are located, generally, in areas of little pronounced reliefs. In fact, a large part of the region is a plain, a plateau, which constitutes the region called La Mancha. In the center of La Mancha, there is an area that borders four provinces (Albacete, Ciudad Real, Cuenca and Toledo) and is called "La Mancha Húmeda." It is located at the Upper Guadiana River Basin and the southern area of the southern plateau or sub-plateau. Wetlands of this area can be classified as

Figure 3.
Location of Castilla- La Mancha region at Spain and Europe. Cartography Service. Junta CLM.

endorheic lagoons, outcrops due to contributions from the underground aquifer and overflowing rivers (tablas)[1] (**Figure 3**) [12, 13].

Some studies define lagoons of endorheic origin as: "... slight depressions in the ground that receive runoff water from the surrounding land during the rainy season, spring and autumn, sometimes very abundant, with torrential episodes affecting small areas. Its permanence depends on depth and environmental temperature of each season". A short and precise definition with which it is agreed [7, 14].

La Mancha is the widest and most perfect plain in the entire Iberian Peninsula. It is a sedimentary basin in which the Guadiana River and its tributaries have not been embedded. Its altitude oscillates between 600 and 700 meters. It has a triangular shape: with a vertex in Ciudad Real capital. It opens toward the east, extending through the provinces of Albacete, Toledo and Cuenca [15].

Rivers that cross these areas do so already in their middle course. When they were, flowing their course was slow and lazy, zigzagging and rambling, which caused the surrounding land to flood due to overflow, since the channels were shallow and had to evacuate large volumes of water in a timely manner, coinciding with moments of copious rain sudden and brief. The water, which was circulating at low speed through the beds of these rivers, gradually were infiltrated into the ground, recharging the aquifers. In some sections of the route, due to the existence of torcas and sinkholes, surface water recharged the underground aquifers. Today they still perform their functions... when there is water [7, 11, 16].

---

[1] The Spanish word "tablas" appears numerous times throughout the text of this chapter, related to wetlands. At La Mancha Húmeda (Central Spain) terrain is very flat, rivers are in their middle course, are shallow and meandering. When the flow increases due to sudden rains, riverbeds are not able to evacuate and overflows occur (flood plains, "tablas"). River seems to widen and flow so slowly that water seems to stop, giving the image of a great plain of still water, so flat as a table or a board. Waters of these overflows usually remain for several weeks, creating a specific ecosystem of great ornithological richness. The best example in Spain was the Tablas de Daimiel National Park, currently with enormous problems to survive as an ecosystem.

There were also outcroppings of subsoil water, called in the area "eyes" which, by themselves, formed lagoons, collaborated in their genesis and made contributions to the fluvial "tablas". They have been left off the hook due to the overexploitation of aquifers for agricultural irrigation [11, 14, 17].

2.2 The weather of La Mancha Húmeda

The Autonomous Community of Castilla-La Mancha presents a climate of contrasts, especially in its central zone, where La Mancha is located.

The climate of La Mancha Húmeda is continental and Mediterranean, characterized by an average rainfall of about 400 mm per year, with a record of temperatures that vary from −15 to 40°C, depending on the time of year [12, 13].

La Mancha is the largest plain in Spain. In contrast to secular dryland crops, it is an area with a relatively high number of endorheic lagoons (many of which have disappeared or are not operational), others marginal to the riverbeds of the rivers that cross the plain, as well as fluvial "tablas" in "La Junta de los ríos" Záncara and Gigüela, Tablas de Villarta and Arenas de San Juan, Tablas de Villarrubia de los Ojos and Tablas de Daimiel [14, 15, 17, 18].

This area marks low rainfall, between 300 and 500 mm/year. Without a defined periodicity, local storms with heavy punctual precipitation occur. They come to cause flooding and significant damage to populations, crops and lagoons in the area. The slope of the land, almost imperceptible, favors the permanence of the waters in the form of lagoons and "tablas."

3. The international importance of La Mancha Húmeda

3.1 Recognitions and figures of protection

The international interest on the wetlands of La Mancha Húmeda has always been motivated by the functions they fulfill in the migration and wintering of european birds, especially for the ducks and waders groups. This is a proven reality [10, 11, 19, 20].

From the first moments, when the European conservationist and scientific associations began to organize and unite (MAR 1960 Project), the Spanish wetlands have been present at the top of the lists of priority areas for conservation. At the 1962 meeting of the MAR Project, in Saintes Marie de la Mer (France), the organizing entities IUCN, ICBP and IWRB declared several Spanish wetlands in Category A. (The reference to the wetlands of Castilla-La Mancha, called Castilla La Nueva in those years, is reproduced verbatim, in the original language).

"ESPAGNE (voir carte n° 8, page 100). CATÉGORIE A … ..4. *Lacs en Nouvelle Castille, Toledo/Ciudad Real a) ca. 39° 00' - 35' N. 3° 30' - 40' W. b) Superficie ca. 1000 ha +. c) 4. 5. 6. 7. d) Nombreux étangs et lacs d'eau douce et alcalins, comprenant les Laguna de Taray, Lagunas de Villafranca, de Alcazar, de Daimiel y Tablas des Giguela, importantes au point de vue Ornithologique. e) Importante région de nidification pour les oiseaux:* Netta rufina *(+ 1000 couples),* Ardea purpurea, Glareola pratincola, Anas platyrhynchos *et quelques* Aythya nyroca. *Quartier d'hiver pour de nombreux Anatidés et* Fulica atra." [20].

La Mancha Húmeda was declared a Biosphere Reserve by UNESCO in February 1981. The numerous and continuous attacks on the wetlands, including Las Tablas de Daimiel National Park itself, led in 2007 to the proposal to declassify it as a Reserve. of the Biosphere, which was halted based on some commitments, partially fulfilled, although they have not given the expected results.

In the list of the Ramsar Convention for the year 2023, there are 76 Spanish wetlands registered, of which 8 belong to Castilla-La Mancha (**Figures 4** and **5**).

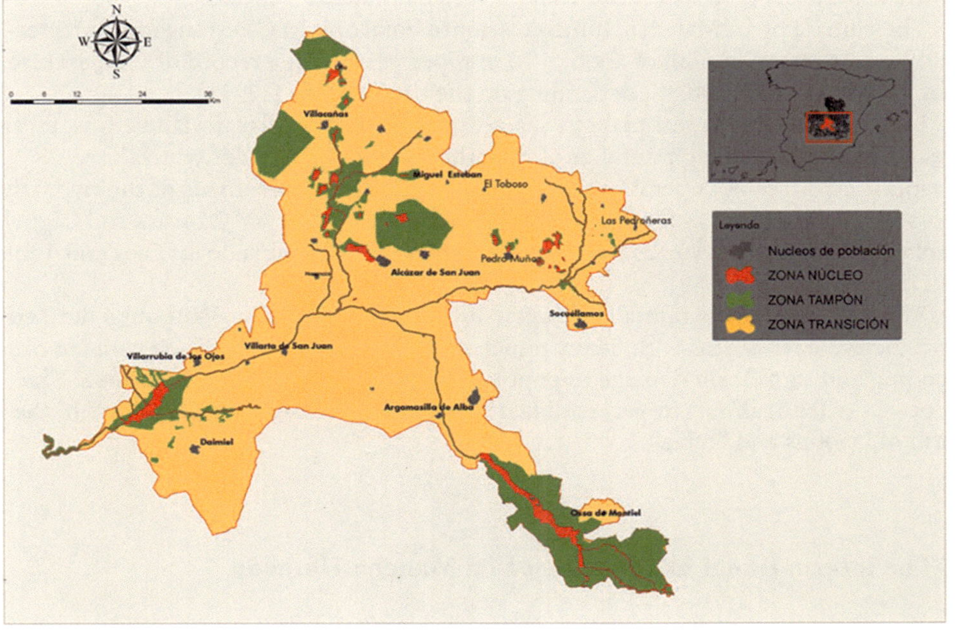

Figure 4.
Location of the La Mancha Húmeda Biosphere Reserve. Cartography Service. Junta CLM.

Figure 5.
Red-crested pochard, male (Netta rufina). Photo T. Palacios.

3.2 The war of quantities

The wetlands extension of La Mancha Húmeda has been the subject of various calculations, "grosso modo," none of them match exactly. In addition to being for self-serving reasons in some cases, in others the variations in the figures could be due to the different areas visited or treated, and not explained in detail in the reports.

The Law of July 17th, 1956 on Cleaning up and colonization of swampy lands that extend immediately to the banks of the Guadiana, Cigüela, Záncara rivers and tributaries of the latter two in the provinces of Ciudad Real, Toledo and Cuenca, it refers in its Preamble "...with an approximate extension of 30,000 Ha. ...and that, through the execution of channeling works and the complementary desiccation, could be rescued for agricultural cultivation and destined to carry out colonization work in them..." [8].

Measurement calculations of the Ministry of Public Works itself, cited in its Projects for the drying of "unproductive" land in favor of agriculture, has been considered valid. It does not seem that this Organism, traditionally contrary to conservationist theses, was going to give incorrect figures on the extension of the land to be "recovered," at least not upwards. The destructive fever, translated into drainage and desiccation, had endorsed the idea that flooding was unhealthy and unproductive areas that had to be cleaned up and recovered.

If, in 1956, 30,000 hectares of wetlands were estimated in different sections of the rivers and their surroundings, it does not seem unreasonable to calculate that there could have been more than 50,000 hectares that still existed in the 19th century.

The studies of this area, carried out in 1972 [21, 22], mention the existence of 20,000 Ha. (of vegas, tablas and flood plains). These areas stretched along the Riansares and Gigüela Rivers, highlighting for their extension the vegas and plains that were defined after the confluence of both rivers and that extended through Herencia, Villarta de San Juan, Arenas de San Juan and Villarrubia of the Eyes until reaching the Tablas of Daimiel [19]. In addition to the Tablas de Daimiel, a large concentration of wetlands of diverse types and genesis was located along this route: lagoons, extensive overflows of "tablas" rivers, upwellings, flooded lands, etc. [19, 21, 22], so it could be considered a mere rough estimate [7].

On November 29, 1994, in the interpellation of a Senator in the Plenary Session of the Senate, the figure of 32,000 Ha was cited, not having been refuted in the response of the Minister of Agriculture, "about the policy measures. general and environmental that the Government will adopt as a matter of urgency to repair and seek solutions to the ecological deterioration that is affecting the Tablas de Daimiel National Park (Ciudad Real)" [23, 24].

3.3 The biosphere reserve of La Mancha Húmeda

La Mancha Húmeda was declared a Biosphere Reserve by UNESCO in February 1981, protecting a total area of 25,000 Ha of wetlands, in which a "fundamental" area of 6000 Ha is located. The numerous and continuous attacks to the wetlands, including the Las Tablas de Daimiel National Park itself, led to the proposal for declassification as a Biosphere Reserve in 2007. This proposal was paralyzed based on some commitments, partially fulfilled, although they have not given the expected results in terms of improving lagoons and the National Park [10, 11, 25, 26].

Figure 6.
Black-tailed Godwit (Limosa limosa) Photo F. Ochoa.

On June 12, 2014, the extension of the Biosphere Reserve was extended to 418,066.16 hectares. This enormous increase in area is due to the inclusion of "buffer zones," "transition zones," population centers, 3 ZEPA areas, 3 LIC, etc., without there being a real increase in the surface area of wetlands or effective actions for their conservation/recovery. "Paper holds everything," the Biosphere Reserve is being expanded and it gives the impression that the surface of wetlands has increased when what has been done has been to include the complete municipal terms of those that have lagoons, ponds or wetlands regardless of their conservation status (**Figure 6**) [27, 28].

In the list of the Ramsar Convention for the year 2023, there are 76 Spanish wetlands registered, of which 8 belong to Castilla-La Mancha: Las Tablas de Daimiel National Park (Daimiel, Ciudad Real), Manjavacas lagoon (Mota del Cuervo, Cuenca), Lagoons de las Yeguas and Camino de Villafranca (Alcázar de San Juan, Ciudad Real), Lagoon of La Vega or del Pueblo (Pedro Muñoz, Ciudad Real), Lagoons of Puebla de Beleña (Guadalajara), Lagoons of Ruidera (Albacete and Ciudad Real), Prado lagoon (Pozuelo de Calatrava, Ciudad Real) and El Hito lagoon (El Hito, Cuenca) (**Figure 7**) [23, 24, 29, 30].

None of the 8 Ramsar sites, located in Castilla-La Mancha, are in an acceptable state of conservation: lack of water supply, clogging of the basin, planting of inappropriate plant species, appearance of invaders, insufficient or no wastewater supply, purified; beds contaminated by heavy metals, drainage ditches, pressure from tourism and leisure activities, very close aerodromes, etc.

4. Loss and degradation of wetlands at La Mancha Húmeda

4.1 General features

The actions on the Spanish wetlands and, specifically, on La Mancha Húmeda (what in current times we would call "environmental impacts"), have their origin

Figure 7.
Marsh Sandpiper (Tringa stagnatillis) feeding on a recent water pond. Photo F. Ochoa.

from previous centuries, not only from the sixties. Drainage and channeling had already been carried out previously in order to prevent water from accumulating in "still" pools of water, the reason for the diseases and fevers that devastated entire populations (the role of mosquitoes was not yet known) transmission of malaria). Another reason for those past canalizations was to avoid flooding in times of flooding. Dozens of deaths occurred due to drowning, when the houses were flooded and crushed due to the collapse of the houses, since the water demolished the walls because they were built of tapial and adobes (blocks of mud and straw), like the vast majority of the houses in the time and area [1–4].

In the last two centuries, the conversion of wetlands into agricultural land was widespread and incessant, throughout Spain. The actions carried out caused the loss, forever, of the "tablas" of La Mancha Húmeda, of the great lagoons of La Janda, Antela and Mar de Campos in northwestern Spain, and of a large part of the Guadalquivir marshes (Doñana), etc. [23]. The loss of wetlands in Spain between 1948 and 1990 has been calculated at 60% of those existing at the time [10].

It may seem that they are only talking about times past, or give the impression that with those enormous losses of wetlands, the fever for their destruction has ended. In La Mancha that never stopped ... continue.

One hundred hectares, more or less, should not distract us from the lack of interest and responsibility that has been shown by the different Spanish organizations with competencies in the conservation of wetlands, deriving and prioritizing interests in only one direction: that of crops, agriculture, encouraging the extraction of water from aquifers with aid from the autonomous governments. Crops that are unthinkable in La Mancha have been planted such as corn, alfalfa and beetroot, which need enormous volumes of water, and must be irrigated by sprinklers, with higher temperatures in summer at 30–35°C, evaporating a large percentage.

With great sadness, we have witnessed, for several years, these risks of waste on the same border with the Las Tablas de Daimiel National Park. Park was dry and the sprinklers watering non-stop. The owners of these crops came to install a plate on the

Figure 8.
Rubble dumped into a lagoon in 1990's. Photo L. Basanta.

sprinklers that bordered the National Park, so that when they turned, none of the water would reach the Park land.

At the same time, wastewaters from any population near a lagoon have been dumped into the glass, as well as solid waste and even industrial wastewaters, with its load of heavy metals, oils and various chemical products (**Figure 8**).

So, for too many years...

4.2 Agriculture: Priority of the destination of water. The irrigation

Let us focus on the recent past and the present: "the greatest pressures have been due, above all, to the intensification of agricultural activity: wetlands are no longer actively dried in a competition for soil resources, but water contributions are withdrawn that feed them in a competition for water resources; and also due to industrial development, infrastructure construction and urban developments related to the tourism sector" [31].

The Ministry for the Ecological Transition and the Demographic Challenge, in its "Strategic Plan for Wetlands to 2030" carries out, with objectivity and sincerity unusual in official publications, an in-depth diagnosis of the situation (2022) of the Spanish heritage of wetlands: "Up to practically In the 70s of the last century, the resource primarily demanded by the agricultural sector was soil, the availability of fertile and productive land, and for this reason many of our wetlands were drained and occupied, actions promoted by the legislative framework and subsidized by the administrations public. Currently it is the water resource" [10].

Before 1970, there were already irrigation systems in Spain that used water from subsoil aquifers. La Mancha was no exception. But those waters were obtained from shallow- or medium-depth wells, conditioned by the drilling machinery of that time.

With limited flows, in turn, by the extraction motors that had to be exterior. The improvement of drilling equipment and, above all, the appearance of the submersible extractor pump, made the fever unleashed by irrigating and planting, the more the better, with subsoil water unbridled.

In 10 years, more than 130,000 hectares were irrigated, which meant an increase of four times the area traditionally irrigated until then [32].

In 1974, 152 hm^3 was extracted and in 1987 it was 600 hm^3. There was a drop in the water level, which implied the need to deepen most of the wells, the depletion of some and a long series of environmental implications causing the disappearance of surface water courses, the deterioration of water quality groundwater (infiltration of water drained from irrigated crops with high salt and nitrate content), the decrease in the degree of humidity of dryland soils and the disappearance of a large number of lagoons that made up the so-called Wet Spot [33].

The "New Water Law" 29/1985, of August 2, contained 9 transitory provisions and 7 additional ones. The transition period from the repealed Water Law of June 13, 1879, to the new one (3 years) was used to request, change, transfer, sell, titles, authorizations, etc., to extract water from aquifers.

During this period, the Basin Agency (Guadiana Hydrographic Confederation) received and authorized concessions for volumes that did not correspond to reality and were much higher than the real recharge capacity of the aquifer [34].

The situation spilled out of control: to the 16,000 legal catchments, we must add 22,000 unauthorized, illegal catchments, according to sources from the Guadiana Hydrographic Confederation itself [34].

The European Water Directive obliges river basin districts to act against the unrestrained pumping of groundwater masses, and sets a time limit: 2027 [33].

In 1985, the alarm had already been raised about the Las Tablas de Daimiel National Park [35]. It was not until February 1987 that the western La Mancha aquifer was declared overexploited [34].

Politicians from any party, neither national nor regional, have dared to put order in the uncontrolled opening of wells without a license, in the chaos of the excessive consumption of unauthorized flows.

Governments have not done enough to shut down illegal wells, fine offenders, and restore the ecological damage caused. Social protests, often violent, by farmers (See Section 4.3), scarcity of "water police," sanctions for offenders are manifestly low...

There is much to do... (**Figures 9** and **10**).

4.3 Unfair actions derived from agricultural practices in wetlands

In full swing of protests by farmers over the declaration of overexploitation of aquifer 23 (Western Mancha), in February 1987, farmers in the affected areas held various assemblies to discuss their problems. In one of the meetings, held in Daimiel, a phrase was pronounced, well remembered by the few conservationists of that time: "water while there is water and then... the donkeys in the shade," which seemed to express the general meaning of those assemblies (let us use the water until it runs out) without an iota of concern for the affected wetlands, nor for the one of greatest regional importance, located in the same municipal area: the Las Tablas de Daimiel National Park.

In March of the year 1988, one of the saddest events that we have experienced in the territory of La Mancha Húmeda took place. It happened during the filming of a series for

Figure 9.
Drainage ditches at the Turleque lagoon. 1988. Photo L. Basanta.

Figure 10.
Drainage ditch at the Camino de Villafranca lagoon. 1988. Photo L. Basanta.

Spanish Television entitled "España Húmeda." On the 23rd we met with the management and the conservation team of the National Park in the Visitor Center, to schedule the day of the following day in which we were going to shoot inside the wetland.

Figure 11.
Las Tablas de Daimiel National Park on fire. March,24th, 1988. Photo L. Basanta.

On March 24, at dawn, from the first high point of the road that leads to the National Park from the town of Daimiel, a dense cloud of smoke was observed that seemed to come from the very heart of the wetland. Unfortunately, it was so: the Daimiel Tablas were on fire!

It seems that some irresponsible people, informed of the presence of our Spanish Television team, tried, with this vandalistic, criminal act, to pressure and publicize their protests and demands. If you are capable of burning a National Park, you are capable of anything (**Figure 11**).

We did not transfer any images for news programs. The broadcast of the Series and the chapter on the National Park took place the following year, between January 3 and March 7, 1989.

It has already been mentioned above that there have been more than 22,000 illegal wells dug in La Mancha. No control of any kind. We have data, based on satellite images, which allow us to affirm that the extractions from the wells considered legal have also exceeded the capacities outlined in their authorizations. They have extracted much more water than authorized.

Another situation detected since the 1980s is that of the occupation of wetlands "by neglect." It is a ladino practice that begins at the time when the shores, or even the entire wetland, are dry. The procedure consists of plowing the shores a few meters around the lagoon, year after year, little by little, each year a little more, until the "useful" surface of the lagoon is reduced, finally plowing the entire lake basin, literally erasing it, especially small- and medium-sized gaps [7, 10].

It is necessary to take into much greater consideration another silent problem that affects endorheic lagoons: Most of them are located in areas with a very little slope so that, due to runoff, it is not easy for them to receive sufficient contributions through scarce rains, because the streams do not begin to flow until the base soil of these small channels is soaked, except for specific cases of local downpours that discharge in their area of influence. These small channels have been modified or canceled in many cases. On this particularity are added the "managements" that some farmers carry out at the time of tillage, destroying or diverting the small and almost imperceptible channels, practically dry for many years, to level their land and gain a few square meters for cultivation [7].

Agricultural activities of diversion or elimination of small channels that flow into rivers or endorheic lagoons must be controlled. It is necessary to respect and recover them, in case they have been intercepted or destroyed.

4.4 Contamination by discharges

During the last three decades of the last century, and before, any lagoon close to a population was used as a dumping ground for contaminated water or solid waste. The less fortunate also received wastewater from the industrial estates: heavy metals deposited in the mud at the bottom, coming from paints and varnishes from door and furniture factories and other activities; debris and sewage in all of them. They were not treated in any way. They were just stored (**Figure 12**).

Several of these lagoons had (and some still have) a level outfall, so that when the levels of water retained in the lagoon were excessive and threatened to affect cultivated land, the excess volume was evacuated directly, through a buried pipe that connected with the Riánsares or Gigüela Rivers, depending on the case, carrying the contaminated waters to the El Taray lagoon or to the Las Tablas de Daimiel National Park itself, downstream.

It coincided that these lagoons were, and still are, sources of botulism. Botulism foci have not disappeared from La Mancha Húmeda. In some wetlands, they are more vigilant than in others and proceed to dry out the lagoon as soon as they see the first affected birds, sending the effluent with some kind of pretreatment, in the best of cases, but always toward Daimiel.

Figure 12.
Farmer irrigating with untreated wastewater, 1985. Photo L. Basanta.

4.5 Natural causes

In the territory of La Mancha Húmeda, the slight slope of the plateau contributes to a natural process in Geography such as the clogging of depressed areas in the endorheic lagoon areas, enormously sensitive to these processes, which ends with the disappearance of the lagoon as such, due to not being able to retain the water or because it does not reach the lagoon. In La Mancha Húmeda, there are striking examples of a phase close to disappearing as lagoons. For example, the Longar lagoon, in Lillo (Toledo), or the Manjavacas lagoon in Mota del Cuervo (Cuenca), etc. [7].

Few more natural causes can be mentioned, in which the human being has not intervened directly or indirectly.

5. Recovery and management of wetlands

The appearance of regional conservation associations at the end of the 1970–80s, and the increase in sensitivity toward wetlands on the part of the Regional Government of Castilla-La Mancha, led to the development of the first catalog of wetlands in the Autonomous Community, with the title "Catalog and Study on the Evolution and Ecology of the Wetlands of Castilla-La Mancha. Spain" 1989 [7].

Until then, wetlands only received the attention of waterfowl hunting enthusiasts. Lagoons such as La Albuera or El Escoplillo, in the municipality of Daimiel, very close to the National Park, are almost mythical names, which witnessed the numerous populations of aquatic birds that exchanged between it and the lagoons in its immediate surroundings and more far away, before they dried up due to being left hanging and without contact with the waters of the aquifer. They were run by the Alcyon Hunters Club (**Figure 13**).

Figure 13.
Spoonbill flock (Platalea leucorodia). Photo A. Amor.

These same hunters recovered more than 2000 hectares of floodplains, turning them into artificial lagoons marginal to the Gigüela River, in the municipality of Quero, with strong private economic outlays in their construction and for hunting purposes, yes. It must be remembered that the first naturalist scientists in advanced countries were hunters. Notable figures in our conservation history were, too.

Few of these artificial lagoons remain operational. When Daimiel began to have serious problems due to lack of water, the Administration's eyes were fixed on the private marginal lagoons of Quero, full of water. The year 1981 began when the first private lagoon was drained by deepening the channel of the Gigüela River and eliminating the water intake. It was the Arroyo Morón Lagoon, in which Biological Station we worked since 1976, until its desiccation.

In 1984, under the "disguise" of cleaning the Gigüela River bed so that the water from the transfer from the Tagus River reached the National Park through its bed, the water intakes of the El Taray lagoons were eliminated, El Masegar, Molino del Abogado, Vega de Quero, Vadoancho, Pastrana and Presarrubias.

The transferred water from the Tagus River barely reached the Park. The very bed of the Gigüela River, with numerous dolinas, filtered the water sending it to the aquifer.

The only wetlands in La Mancha Húmeda that conserved clean water, offered shelter to birds and allowed their reproduction to be successful had been eliminated. Our suggestion that the artificial lagoons of Quero were expropriated by the State, that hunting was prohibited in them, and that they were not drained, was ignored.

Of the "artificial" lagoons from that period, only El Taray remains and, much less extensive than the original, in favorable years, El Masegar.

El Taray lagoon is an artificial, private lagoon. An example of good management and conservation. Hunting has not been done there for many years and photography is now allowed from observatories (**Figure 14**).

El Taray lagoon is maintained with water from the Riánsares River. Originally, they were "tablas" formed by the "junta" of the Gigüela and Riánsares Rivers.

Figure 14.
Black-winged stilt (Himantopus himantopus). Photo F. Ochoa.

Dams were built more than a century ago to retain water and raise Carp (Cyprinus carpio) for sale. Later, when dedicating itself to hunting, more dikes (cespederas) were arranged to better manage the water. A dam was built to separate a part of El Taray, which was called El Masegar, that was sold in its day and managed independently.

A conscientious study has concluded that the El Masegar lagoon and others in its immediate surroundings can be recovered if the natural functioning of the Gigüela River is restored, especially affected by the excavation and channeling of its channel [14]. This lagoon belonged to the José María Blanc Foundation, an entity of which we have been a part for several years, carrying out various studies on wetlands of La Mancha and Doñana.

Another artificial lagoon, privately owned, is the Dehesa Monreal lagoon, at the municipality of Dos Barrios (Toledo). Built with all the licenses, it is an example of good conservation management, although it was created at the end of the 1970s for the purpose of hunting. It is located on the banks of the Arroyo Cedrón. It has played a fundamental role in the recovery of Malvasía (Oxyura leucocephala) in Spain.

A municipal initiative that should be highlighted is the one that the City Council of La Guardia (Toledo) undertook in 2013 and 2015, acting on the reedbed adjacent to the Arroyo Cedrón (upstream of the Dehesa Monreal lagoons), to build three lagoons that have favored several species of endangered marsh birds, such as Bearded tit (Panurus biarmicus), Reed bunting (Emberiza schoeniclus) and Mustached warbler (Acrocephalus melanopogon), in addition to constituting one of the largest roosts in Spain for the Marsh harrier (*Circus aeroginosus*) with more than 250 specimens together (**Figure 15**).

In the most recent years, a certain interest in the recovery of some lost lagoons has begun. The interests are usually touristic, to attract visitors to the nearest city or town.

Figure 15.
Drone view. La Guardia (Toledo) lagoons. An initiative of the City Council and the Regional Government. Photo M. Guzmán.

The so-called "recoveries" are not designed according to biological criteria: ornamental plant plantations are carried out and picnic tables are installed. Plant species such as Tamarix, a great detractor of water and an invasive species, have been introduced. The advisers, if there are any, have not known these wetlands in their best moments and repeat not very recommendable models of other equally incorrect recent actions.

A lagoon from La Mancha in 1972 did not have a row of tamarisk trees surrounding its perimeter. The lagoons of La Mancha, in general, did not have arboreal vegetation. In the recoveries, on many occasions, examples of design of artificial lagoons for hunting purposes have been followed.

We do not want to give here the names of affected lagoons. This is probably not the place to do it.

6. Current situation of wetlands at Castilla-La Mancha

In Castilla-La Mancha, it would be said that they have used their rivers for experimentation and example of drying theories. Thus, there is currently not a single important channel that crosses the plain of La Mancha that has not been "managed" by man.

Rivers Záncara, Gigüela, Guadiana, Valdespino, etc., have seen their sinuous paths become rectilinear channels.

The ecological disaster of these actions has been enormous: Several thousand hectares of rich marsh areas have disappeared forever and with them the animals that inhabited them, in such quantity and variety that they deserved international attention.

Lagoons have followed the same path and method. Those that were not filled with solid waste received wastewater and industrial discharges. In other cases, a ditch in its channel ended the life expectancy of aquatic fauna and flora.

Wetlands that survive in La Mancha go through greater difficulties every year. Those that are not dry are because they receive wastewater.

In the list of the Ramsar Convention for the year 2023, there are 76 Spanish wetlands registered, of which 8 belong to Castilla-La Mancha. This is the current situation.

- Las Tablas de Daimiel National Park: Its natural functioning has been destroyed. There has been no "junta" of the waters of the Guadiana and Gigüela Rivers for more than 40 years. The flow of these rivers is a testimonial. The main cause of the limited state in which the National Park is located is the drop in the water level caused by the massive extraction of water from the aquifers for agricultural irrigation. Of the wells drilled, 2/3 are illegal, without a license. The National Park, with enormous difficulties, survives thanks to the work of guards and technicians, the transfers it receives from the Tajo River and the punctual contribution of water pumped from emergency wells, which allows having a minimum extension of open water to avoid the spontaneous combustion of peat, which is the main substrate of the National Park.

- Laguna de Manjavacas (Mota del Cuervo. Cuenca): Traditionally it has been the best lagoon in La Mancha Húmeda to observe migrating waders, and, depending on the water level, to establish breeding colonies of Black-necked grebes

(Podiceps nigricollis) and various species of larolimicolae. It is in the final phase of clogging, accelerated by the continuous contribution of wastewater from the population of Mota del Cuervo for decades. It was a typical La Mancha endorheic lagoon, saline, with specialist vegetation on the shores (*Suaeda sp. Salicornia sp.* etc). For 30 years a reedbed has developed around the point of entry of wastewater, which undermines its original character. The Greater Flamingo (Phoenicopterus roseus) has come to reproduce. We suspect that these colony nestings have been induced by unscrupulous amateurs, who have built nest starts to provoke acceptance by the birds. Another serious mistake, if it had been so.

- Lagoons de las Yeguas and Camino de Villafranca (Alcázar de San Juan. Ciudad Real). Laguna de las Yeguas is hypersaline. Its state of conservation is relatively good. It is in danger of receiving contaminated water, as has already happened because it is connected to the Camino de Villafranca lagoon by means of a canal. This last lagoon has been the one that has received, for decades, untreated wastewater from the town of Alcázar de San Juan and its industrial park. These waters were so poisonous that in a few minutes, they rusted a piece of metal that was introduced into them. It has been completely emptied on different occasions. The contaminated water has been conducted through the La Veguilla canal to the Gigüela River and directly to the Las Tablas de Daimiel National Park. It currently receives water with a certain degree of treatment from the Alcázar de San Juan Treatment Plant, which is discharged right at the outlet of the treatment plant, forming the La Veguilla lagoon. They have always been dangerous lagoons because at the end of the summer, the botulism epidemic has always developed, with significant mortality of birds. In recent times, it has been decided to empty them of water as soon as the first dead birds appear and the water from the treatment plant is sent directly to the Gigüela riverbed.

- Laguna de la Vega or del Pueblo (Pedro Muñoz. Ciudad Real). Lagoon of endorheic origin is highly polluted by contributions of wastewater from the village. It has always been a privileged lagoon for the reproduction of colonies of Black-necked Grebes and White-faced Terns (Chlidonias hybrida), Ducks and Common Pochards (Aythya ferina) and Black-headed Gulls, (Chroicocephalus ridibundus) despite the fact that it has always been affected by botulism. Currently, it dries up in some seasons to avoid mortality. The density of birds in the breeding season is impressive.

- Lagoons of Puebla de Beleña (Puebla de Beleña, Guadalajara). Both are not located in La Mancha. In other times, with water, it housed an important colony of Black-necked grebes. Currently, the water level is at −1.20 m, according to two witness wells. Its state of conservation is very good, except for the lack of water and the proximity of an airfield.

- Ruidera Lagoons (Ruidera, Argamasilla de Alba, Alhambra and Villahermosa. Ciudad Real) (Ossa de Montiel. Albacete). There are a total of 15 interconnected lagoons, through which the water that communicates them runs in stages. Those located at a lower altitude are the most interesting for birds. It is an area that suffers strong tourist and real estate pressure. The volumes of water have decreased notably due to irrigation located at the highest points.

- Prado Lagoon (Pozuelo de Calatrava. Ciudad Real). It is of pseudokarstic and seasonal origin. It usually dries up in summer. Numerous pairs of Avocet (Recurvirostra avosetta) and Stilt (Himantopus himantopus) reproduce. During the migratory period is visited by waders and in winter by ducks. It is located next to the population.

- Lagoon of El Hito (El Hito and Montalvo. Cuenca). It is an endorheic lagoon of 573 hectares in a good state of conservation. Its greatest importance is hosting large flocks of Cranes (Grus grus) on their autumn and spring migrations.

Spanish Inventory of Wetlands lists as existing lagoons that were dried up when the river intakes were eliminated. As an example, the Arroyo Morón lagoon is one of them. Dry since 1981. Other inaccuracies are due to the double annotation of the same lagoon basin with a different name but identical coordinates. It is a great job but still inaccurate.

The Ministry for the Ecological Transition and the Demographic Challenge has published a short report on the state of Ramsar wetlands in Spain: About 12% of the sites are well preserved, 15% is in a moderate state, 24% are in a poor state, 30% is in a very poor state and 19% are under evaluation.

"Among the inland aquatic environments, floodplains have suffered a greater degree of alteration, being entirely modified by man, followed by freshwater wetlands (62% modified), karsts (54%) and saline (51%). Mountain lakes and wetlands are those with the best state of conservation, as they have been less intervened. On the other hand, more than two thirds of the inventoried coastal wetlands are considered altered or degraded, while those that are conserved do not reach one third" [10].

7. Conclusion

This account of part of the history of the evolution of the wetlands of La Mancha leaves, once again, a strong bitter taste for some conservationists who saw the splendor of the "tablas" of Villarta and Arenas de San Juan in 1972, the "Junta" of the Záncara and Gigüela Rivers, and Las Tablas de Daimiel National Park in 1975, and up to the present 2023.

There has been no respect for wetlands or the National Park nor is there now. Very few concrete actions are carried out. It is almost better that way, because when they do one, are making mistakes of Conservation. Fortunately, nature always puts those who plant tamarisk trees around the shores of Lagoon X in their place, drying up the planted specimens.

Various villages close to lagoons have been equipped with treatment plants, but the water that is discharged into them is not always adequately purified. Sometimes due to breakdowns, other times due to lack of available budget, the treatment plant does not work. The quality of the effluents from the treatment plants is not always optimal, with untreated direct discharges being made into the beds of lagoons and rivers.

Now, the objective seems to be to demonstrate that we have a lot of protected territory. A set of figures and numbers in which the extension of wetlands is mixed with other uses, reaching surfaces that place us in third place in the world in Ramsar sites and first in Unesco Biosphere Reserves. Are really protected?

La Mancha Húmeda situation is so desperate that it calls for desperate measures.: ¡To each lagoon its well!. Conservation must fight with the same weapons as agriculture. The excavation of a well in the vicinity of a lagoon, and the provision of an extractor motor, would allow to control the adequate water levels. We must guarantee water in the lagoons in the most sensitive biological periods. The hatred toward wetlands seems to be carried in the blood and inherited. So many years and it seems that we have not learned anything. We are arriving too late …

Acknowledgements

We must acknowledge the vocational work in favor of La Mancha Húmeda wetlands by: José María Blanc, Francisco León, José Baragaño, Manuel Sáenz de Viedma, Francisco Bernis, Ramón Sáez-Royuela, Ramón Coronado, José Luis Oriol, Javier Castroviejo, Pedro Molina Vicente, Cosme Morillo, Germán García, Carlos Roquero, Ismael Riquelme, Carlos Otero, Fernando Portillo, Juan Rubio, Tom Gullick, Keith Emmerson, José Jiménez, Jesús Casas, Mª Jesús Sánchez, Manuel Carrasco, Carlos Ruiz, Antonio Aranda, Bautista García-Consuegra and his sons Bautista, Jesús and Prisco, Angel Clavería, Alejandro del Moral, Julio Escudero, Francisco Escudero, Hermanos Santos, Enrique Arias, Beltrán Ceballos, Carlos Llandres, Blanca Ramos, Raúl Martínez, Blanca Soto, Luis Rodríguez, José A. Basanta, Eduardo Castilla, Carlos Turón and his sons Carlos and Jorge, Antonio Carmena, Pablo Pereira, Javier Pasamontes, Enrique del Castillo, Macario the guardian of El Taray, Santiago Reino "El Pincha" and many others friends with whom we share birdingdays of waders and ducks, and to whom I apologize for not being reflected here. Thanks also to the photographers who have lent the photos for this chapter.

Author details

Luis Fernando Basanta Reyes
Grupo de Investigación Biología Evolutiva y de la Conservación, Facultad de Ciencias Biológicas, Universidad Complutense de Madrid, Madrid, Spain

*Address all correspondence to: lbasanta@ucm.es

References

[1] González M, J.A. La desecación de humedales en los inicios del siglo XIX: un proyecto de ingenieros militares en La Mancha. Cuadernos de Geografía. 2022;**108-109**(1):149-173 Valencia (España)

[2] Bertola M, Mazzucato M, Pombi M, et al. Updated occurrence and bionomics of potential malaria vectors in Europe: A systematic review (2000–2021). Parasites Vectors. 2022;**15**:88. DOI: 10.1186/s13071-022-05204-y

[3] Tajuelo ML. El Paludismo en España desde una perspectiva histórica. Facultad de Farmacia. Universidad Complutense de Madrid; 2017. Available From: https://eprints.ucm.es/id/eprint/55457/

[4] https://lamalaria.com/malaria/malaria-espana/. [Accessed: January 28, 2023]

[5] Abellán Contreras, F.J. Consideraciones histórico-jurídicas sobre el saneamiento de terrenos pantanosos, lagunas y marismas en España (SS.XIX-XX): Exégesis de la «Ley Cambó». Revista Jurídica de Castilla y León, Núm. 58. Sept. 2022. ISSN: 2254-3805

[6] Gaceta de Madrid. Núm 208, 27 julio 1918. pp. 268-270. Ley Cambó

[7] Basanta Reyes LF. Estudio Inventario sobre la Evolución y la Ecología de las Zonas Húmedas de Castilla-La Mancha (5 vols.). In: Dirección General de Montes, Caza y Pesca. Consejería de Agricultura. Toledo: Junta de Comunidades de Castilla-La Mancha; 1989

[8] Boletín Oficial del Estado. Ley sobre saneamiento y colonización de los terrenos pantanosos que se extienden inmediatos a las márgenes de los ríos Guadiana, Cigüela, Záncara y afluentes de estos dos últimos en las provincias de Ciudad Real, Toledo y Cuenca. Ley de 17 de julio de 1956. Madrid: B.O.E; 1956. Núm. 200. pp. 4693-4695

[9] https://www.ramsar.org/es. [Accessed: February 01, 2023].

[10] MITECO. Situación actual y perspectivas de conservación y restauración de los humedales españoles hasta 2030. Informe. Madrid: Ministerio para la Transición Ecológica y el Reto Demográfico; 2020

[11] Casado, S., Florín, M., Molla, S. & Montes, C. Current status of Spanish Wetlands, en Finlayson, M, Hollis, T. & Davis, T. (Eds) Managing Mediterranean wetlands and their birds. En Comunic. Comisión EU 1994. Actas del Simposio Internacional del IWRB de Grado, Italia, febrero de 1991. Publicación especial del IWRB n° 20, 1992, GB. Istituto Nazionale di Biologia della Selvaggina, Italia. Páginas 56-58.

[12] https://www.castillalamancha.es/clm/geografia. [Accessed: March 2, 2023].

[13] Pérez González ME, Sanz Donaire JJ. Clima y microclima de La Mancha Húmeda. Anales de Geografía de la Universidad Complutense 1998;**18**: 239-256. ISSN: 0211-9803 Madrid. España

[14] Fornés JM, de la Hera A, Cruces J, Llamas MR. Funcionamiento hidrológico de un humedal ribereño, El Masegar, en el conjunto de los humedales de La Mancha Húmeda de la Cuenca Alta del Río Cigüela. Estudios Geológicos. 1995; **51**:259-276 CSIC. Instituto de Geociencias. Madrid

[15] Díaz Moreno JL, Sierra GJ, Vázquez González A, Zárate Martín A, de

Castilla-La A, Mancha. Junta de Edita: Junta de Comunidades de Castilla-La Mancha. Toledo; 1986

[16] Basanta Reyes, L.F. Informe. Había una vez ... Las zonas húmedas. Revista Añil (Cuadernos de Castilla-La Mancha).1993;**2**:13-15. Toledo. https://www.researchgate.net/publication/344954030

[17] García Rodríguez M. Hidrogeología de las Tablas de Daimiel y de los Ojos del Guadiana. In: Bases hidrogeológicas para una clasificación funcional de humedales ribereños. Tesis Doctoral. UCM Madrid: Facultad de CC Geológicas; 1996

[18] López-Rey Lumbreras D. Riadas históricas en la provincia de Toledo. In: Análisis meteorológico e impacto social. Tesis de Lic. Madrid: UCM; 2007

[19] Madrazo B, Francisco. Información española sobre anátidas y fochas (Época invernal). Publicaciones de la Sociedad Española de Ornitología. 1964

[20] Olney PJS, the staff of the IWRB/MAR Bureau. IUCN Publications New Series. N°5. Sponsored by IUCN, ICBP and IWRB; 1965

[21] Club de Cazadores "ALCYON". Nuevos datos sobre la nidificación de Larus ridibundus en España (region manchega). In: Volumen Especial. 459-471. Ardeola, International Journal of Ornithology. Madrid: Sociedad Española de Ornitología; 1971

[22] Bernis Madrazo F, Compte Sart A. Informe sobre las Tablas de Daimiel y otras zonas húmedas manchegas. Informe inédito. 1972

[23] Martín Mateo, R. La protección de las zonas húmedas en el ordenamiento español. Revista de Administración Pública, 1981. pp. 7-31. Núm. 96. Septiembre-diciembre. Madrid.

[24] Senado Cortes Generales. Diario de Sesiones del Senado V Legislatura. 1994. Núm. 56. Interpelación Parque Nacional de las Tablas de Daimiel (Ciudad Real)

[25] https://www.ramsar.org/es/acerca-de/historia-de-la-convencion-sobre-los-humedales. [Accessed: March 7, 2023].

[26] Pérez M, Aramburu MJ, González JL. Reserva de la Biosfera de la Mancha Húmeda: retos y oportunidades de futuro. In: Dirección General de Áreas Protegidas y Biodiversidad. Toledo: JCCM; 2011

[27] https://www.castillalamancha.es/gobierno/desarrollosostenible/dgapfyen/actuaciones/reserva-de-la-biosfera-de-la-mancha-humeda [Accessed: February 2, 2023]

[28] Barbier EB, Acreman MC, Knowler D. Valoración económica de los humedales – Guía para decisores y planificadores. Gland, Suiza: Oficina de la Convención de Ramsar; 1997

[29] UNESCO. Convention on Wetlands of International Importance Especially as Waterfowl Habitat. 1994. Irán: Ramsar; 1971

[30] SEO/BirdLife. In: SEO/BirdLife, editor. Humedales Ramsar en España: En Alerta. Madrid: SEO/BirdLife; 2022

[31] Viñals MJ, Blasco D, Morant M, editors. Los humedales mediterráneos: el contexto ambiental y social. In: Reflexiones para su estudio y gestión eficaz. Fundación Biodiversidad de la presente edición Editorial Universitat Politècnica de València; 2011. p. 266

[32] Sancho Comins J, García-Abad Alonso JJ, Navalpotro Jiménez P,

Santaolalla A, Martínez Vega J. La tradición e innovación en el paisaje agrario: los efectos de la PAC en la región central española. El campo: boletín de inf ormación agraria, 1994. pp. 215-234. ISSN 0212-2146

[33] Rosell J, Viladomiu L. El Programa de Compensación de Rentas por reducción de regadíos en Mancha Occidental y Campo de Montiel. Miteco: Economía Agraria, n.° 179; 1997. pp. 331-350

[34] WWF/ADENA. 2006 Uso ilegal del agua en España. Causas, efectos y soluciones. info@wwf.es. [Accessed: February 4, 2023]

[35] Basanta Reyes LF. Tablas de Daimiel: SOS. Castilla-La Mancha. Revista de Información de la Junta de Comunidades. 1985;**Julio N° 17**:14-15 Toledo

Chapter 5

Bird Diversity Changes in Tervasaari, Savonranta, Finland from 1989 to 2023

Heimo Mikkola and Anita Mikkola

Abstract

Birds are sensitive to climate and environmental changes and monitoring the abundance and distribution of bird populations can be excellent barometers of the health of the environment. To investigate bird diversity changes, a hobby like birding was done from May 1989 until July 2023 on a forested island, at Lake Orivesi in Eastern Finland. The main observation area was some 1 km to all directions from the house mainly used for summer holidays. A total of 145 bird species were identified during the 34 years, 108 during the first period 1989–2000, 102 during the second 2001–2013 and 114 during the third period 2014–2023. Abundance classification listed 48% of species as common, 20% as rare and 32% as occasional. The population trend for 80% of bird species seen more than once or one year in 17% of species have upward trend, in 16% of species trend is downwards and the remaining 67% of species are not showing any clear changes. Bird feeding attracted 28% of species mainly during the winter but later also almost all year around. Interesting behaviour changes have been noted at individual and species levels during the feeding activity and in the use of the nest boxes.

Keywords: long-term birding, climate change, bird feeding, the behaviour of birds, fight for nest boxes

1. Introduction

Distribution ranges of bird species around the globe are expected to contract in response to climate change. The first study addressing the potential consequences of climate change on the biology of birds dates back 32 years [1]. Since then we have gained tremendous knowledge in this field, and every week provides large amounts of information through a steady stream of new publications [2, 3]. A lot of attention has been paid to possible climate change impacts on bird diversity but often with very short observation periods or only covering part of the species [4, 5].

One can read extremely alarming news about the state of the World's birds according to which one in eight bird species is threatened with extinction [6]. Similar news from Europe indicates that whilst 47% of 463 bird species in the European Union (EU) are in good conservation status, 39% are in poor and bad conservation status [7]. The fact that 14% of all EU bird species have an unknown status highlights the need for this type of study [8].

We have done hobby-like birding on a forested island, at Lake Orivesi in Eastern Finland from May 1989 until July 2023. This long-term data set makes it possible to explore the bird diversity changes in that fairly remote area. The main observation area was some 1 km to all directions from the house mainly used for summer holidays. This paper gives also several particular behaviours of different bird species, especially those attending our bird feeding activities or using the offered nest boxes.

2. Material and methods

Since 1989, we have had a forest house in Tervasaari, Savonranta, 65 km south of Joensuu and 70 km north of Savonlinna. Tervasaari is a small island at Lake Orivesi in East Finland connected with a narrow road to the mainland. Over the years, we have marked down most birds we have seen in the area, which is some 1 km to all directions from the house. Our land includes 600 m of lakeshore and some 3 ha of forest. First years we spent only the summer holidays in that area while working abroad and now after 2007 also winter-time birding has been possible. This however did not affect much in the total bird numbers, as most of the species are migratory.

Tervasaari has been occupied by humans at least for well over 300 years but now only two houses are occupied a year around and two other houses are for summer use only. Until 1900, slash-and-burn method was used in farming and tree burning for tar gave even the name for the place (Tervasaari = tar island). Also later the forest cover has been harvested to some extent but we have avoided

Figure 1.
Unfortunate forest harvesting of the neighbours. Photograph by Heimo Mikkola, Finland.

DOI: http://dx.doi.org/10.5772/intechopen.1002390

falling any large trees from our area. Even the firewood is purchased from outside the area. Small trees were harvested more when we had sheep during one summer. Neighbours are using their forests more effectively (see **Figure 1**), making

Figure 2.
Northern Hawk Owl Surnia ulula. *Photograph by Esko Rajala, Finland.*

Figure 3.
Great Grey Owl Strix nebulosa. *Photograph by Esko Rajala, Finland.*

Figure 4.
Pygmy Owl Glaucidium passerinum. *Photograph by Esko Rajala, Finland.*

Figure 5.
Black Woodpecker Dryocopus martius. *Photograph by Esko Rajala, Finland.*

Figure 6.
White-backed Woodpecker Dendrocopos leucotos. *Photograph by Ari Rantamäki, Finland.*

Figure 7.
Great Spotted Woodpecker Dendrocopos major. *Photograph by Esko Rajala, Finland.*

our forest a haven for some forest species, like owls (Strigidae) (**Figure 2–4**) and woodpeckers (Picidae) (**Figure 5–7**) etc. **Figures 8–15** are illustrating some typical and rare Tervasaari birds.

To reduce the possible inconsistencies in writing down every year's observations from most common birds, a standard number of years for each period was taken from Chaffinch *Fringilla coelebs*, which is the most common bird species in Finland [9]. Against that value, the occurrence index was calculated for each other recorded species. Criterias used in abundance and population trend estimations were as follows: C = Common (7 or more observation years); R = Rare (4–6 observation years) and O = Occasional (1–3 observation years); U = upwards (at least 0.4 up), D = downwards (at least 0.4 down) and N = no major trend up or down (less than

Figure 8.
Whooper Swan Cygnus cygnus. *Photograph by Esko Rajala, Finland.*

Figure 9.
Barnacle Goose Branta leucopsis. *Photograph by Esko Rajala, Finland.*

Figure 10.
Sparrowhawk Accipiter nisus. *Photograph by Esko Rajala, Finland.*

Figure 11.
Common Crane Grus grus. *Photograph by Esko Rajala, Finland.*

Figure 12.
Waxwing Bombycilla garrulus. *Photograph by Esko Rajala, Finland.*

Figure 13.
Red-flanked Bluetail Tarsiger cyanurus. *Photograph by Jari Peltomäki, Finland.*

0.4 up or down), N/A = seen only once or during one year, so no population trend estimation possible (**Table 1**).

Bird feeding was undertaken in all years after 2005. The main food served were groundnuts, sunflower seeds, lard and one kg Tintin fat seed bars (60% of animal fat plus oat, soy, vegetable oil and sunflower seeds). In addition, household food remains were given mainly to the larger birds like crows (Corvidae) and gulls (Laridae). Every

Figure 14.
Great Grey Shrike Lanius excubitor. *Photograph by Esko Rajala, Finland.*

Figure 15.
Pine Grosbeak Pinicola enucleator. *Photograph by Esko Rajala, Finland.*

Number	Species	1989–2000	2001–2013	2014–2023	Status/ Abundance	Trend
1	*Gavia arctica*	0	0.4	0.8	C	U
2	*Podiceps cristatus*	0.5	0	0.2	R	D
3	*Podiceps griseigena*	0.5	0.9	0.8	C	U
4	*Botaurus stellaris*	0.4	0.7	0.7	C	N
5	*Ardea cinerea*	0	0	0.1	O	N/A
6	*Cygnus cygnus*	0.1	1.0	1.0	C	U
	Arctic geese unidentified	0	0.4	0.9	C	U
7	*Anser fabalis*	0	0.1	0.4	R	U
8	*Anser brachyrhynchus*	0	0	0.2	O	N
9	*Anser albifrons*	0	0.1	0.2	O	N
10	*Anser erythropus*	0	0	0.4	R	U
11	*Branta canadensis*	0	0.1	0.1	O	N
12	*Branta leucopsis*	0	0	0.7	R	U
13	*Branta bernicla*	0	0	0.4	R	U
14	*Anas penelope*	0.4	0.4	0.8	C	U
15	*Anas crecca*	0.4	0.2	0.3	C	N
16	*Anas platyrhynchos*	0.4	0.1	0.3	C	N
17	*Anas querquedula*	0.1	0	0	O	N/A
18	*Anas clypeata*	0.3	0	0	O	N
19	*Clangula hyemalis*	0	0	0.2	O	N
20	*Melanitta fusca*	0	0.1	0.1	O	N
21	*Bucephala clangula*	1.1	0.7	0.7	C	D
22	*Mergus serrator*	1.0	0.4	0.5	C	D
23	*Mergus merganser*	0.4	0.3	0.7	C	N
24	*Pernis apivorus*	0.3	0.1	0.2	R	N
25	*Accipiter gentilis*	0.1	0	0.1	O	N
26	*Accipiter nisus**	0.3	0.7	0.8	C	U
27	*Buteo buteo*	0.3	0	0.4	R	N
28	*Aquila clanga*	0.1	0	0.2	O	N
29	*Pandion haliaetus*	0.4	0.2	0.4	C	N
30	*Falco tinnunculus*	1.3	0.2	0.1	C	D
31	*Falco columbarius*	0.1	0	0.1	O	N
32	*Falco subbuteo*	0.6	0.2	0.3	C	D
33	*Falco peregrinus*	0.3	0	0.2	R	N
34	*Tetrastes bonasia*	0.1	0.3	0.5	C	U
35	*Lagopus lagopus*	0	0.1	0	O	N/A
36	*Tetrao tetrix*	1.1	0.3	0.4	C	D

Number	Species	1989–2000	2001–2013	2014–2023	Status/ Abundance	Trend
37	*Tetrao urogallus*	0.3	0.2	0.4	C	N
38	*Phasianus colchicus*	0.1	0	0	O	N/A
39	*Porzana porzana*	0.1	0	0	O	N/A
40	*Fulica atra*	0.1	0	0	O	N/A
41	*Grus grus*	1.1	0.8	0.9	C	N
42	*Vanellus vanellus*	0.4	0.2	0.2	R	N
43	*Lymnocryptes minimus*	0.1	0	0	O	N/A
44	*Gallinago gallinago*	0.8	0.2	0	C	D
45	*Gallinago media*	0.1	0.1	0.1	R	N
46	*Scolopax rusticola*	0.8	0.2	0.8	C	N
47	*Limosa lapponica*	0	0.1	0	O	N/A
48	*Numenius arquata*	0.6	0.8	0.4	C	N
49	*Tringa nebularis*	0.3	0.2	0.1	R	N
50	*Tringa ochropus*	0	0.2	0.2	R	N
51	*Tringa glareola*	0	0	0.1	O	N/A
52	*Actitis hypoleucos*	0.5	0.7	0.5	C	N
53	*Hydrocoloeus minutus*	0.3	0	0.1	R	N
54	*Larus ridibundus*	0.6	0.4	0.2	C	D
55	*Larus canus**	0.9	1.0	1.0	C	N
56	*Larus fuscus*	0.1	0.3	0.2	R	N
57	*Larus argentatus**	0.4	0.7	0.5	C	N
58	*Rissa tridactyla*	0	0	0.1	O	N/A
59	*Sterna hirundo*	0.5	0.3	0.8	C	N
60	*Sterna paradisaea*	0.1	0	0	O	N/A
61	*Chlidonias niger*	0	0	0.2	O	N
62	*Columba oenas*	0	0	0.1	O	N/A
63	*Columba palumbus**	0.4	0.3	0.7	C	N
64	*Cuculus canorus*	0.8	0.7	0.8	C	N
65	*Bubo bubo*	0	0.2	0.1	R	N
66	*Surnia ulula*	0	0	0.1	O	N/A
67	*Claucidium passerinum**	0.1	0.3	0.3	C	N
68	*Strix uralensis**	0.3	0.4	0.2	C	N
69	*Strix nebulosa*	0.1	0.2	0	R	N
70	*Asio otus*	0	0.1	0	O	N/A
71	*Aegolius funereus*	0	0.1	0.1	O	N
72	*Jynx torquilla*	0.1	0	0	O	N/A

Number	Species	1989–2000	2001–2013	2014–2023	Status/ Abundance	Trend
73	*Picus canus**	0	0.2	0.8	C	U
74	*Dryocopus martius**	0.4	0.4	0.9	C	U
75	*Dendrocopos major**	1.3	0.9	1.0	C	N
76	*Dendrocopos leucotos**	0.3	0	0.5	R	N
77	*Dendrocopos minor**	0.1	0.6	0.2	C	N
78	*Picoides tridactylus*	0.1	0.1	0	O	N
79	*Hirundo rustica*	0.4	0.1	0.2	R	N
80	*Delichon urbica*	0.1	0	0.1	O	N
81	*Anthus trivialis*	0.6	0.6	0.2	C	D
82	*Motacilla alba*	1.3	1.0	0.6	C	D
83	*Bombycilla garrulus**	0	0.2	0.5	C	U
84	*Prunella modularis**	0.3	0.2	0.1	R	N
85	*Erithacus rubecula**	1.0	0.3	0.6	C	D
86	*Luscinia luscinia*	0.3	0	0.1	R	N
87	*Tarsiger cyanurus*	0	0.1	0	O	N/A
88	*Phoenicurus ochruros*	0.1	0	0	O	N/A
89	*Phoenicurus phoenicurus*	0.5	0.2	0.1	C	D
90	*Saxicola rubetra*	0.1	0	0	O	N/A
91	*Zoothera dauma*	0	0.1	0	O	N/A
92	*Turdus merula**	0.6	0.9	1.0	C	U
93	*Turdus pilaris**	1.0	0.7	0.8	C	N
94	*Turdus philomelos*	0.8	0.6	0.6	C	N
95	*Turdus iliacus*	0.6	0.9	0.6	C	N
96	*Turdus viscivorus*	0.1	0.1	0.1	R	N
97	*Accrocephalus schoenobaenus*	0.6	0	0.2	R	D
98	*Acrocephalus dumetorum*	0.1	0.1	0	O	N
99	*Hippolais icterina*	0.3	0.3	0	R	N
100	*Sylvia curruca*	0.4	0.4	0.2	C	N
101	*Sylvia communis*	0.1	0	0	O	N/A
102	*Sylvia borin*	0.8	0.6	0.4	C	D
103	*Phylloscopus trochiloides*	0.4	0.2	0	R	D
104	*Phylloscopus sibilatrix*	0.5	0.7	0.7	C	N
105	*Phylloscopus collybita*	0.1	0.1	0.3	R	N

Number	Species	1989–2000	2001–2013	2014–2023	Status/ Abundance	Trend
106	*Phylloscopus trochilus*	1.0	0.6	0.7	C	N
107	*Regulus regulus*	0	0.2	0	O	N
108	*Muscicapa striata*	0.4	0.8	0.4	C	N
109	*Ficedula parva*	0	0	0.1	O	N/A
110	*Ficedula hypoleuca*	1.1	1.0	0.9	C	N
111	*Panurus biarmicus*	0.1	0	0	O	N/A
112	*Aegithalos caudatus**	0	0.3	0.2	R	N
113	*Poecile palustris**	0	0.2	0.1	O	N
114	*Poecile montanus**	0.9	0.9	1.0	C	N
115	*Poecile cinctus*	0.1	0	0	O	N/A
116	*Lophophanes cristatus**	0.4	0.4	0.4	C	N
117	*Periparus ater**	0.4	0.4	0.8	C	U
118	*Cyanistes caeruleus**	1.0	0.9	1.0	C	N
119	*Cyanistes cyanus*	0	0.1	0	O	N/A
120	*Parus major**	0.9	1.2	1.0	C	N
121	*Sitta europaea**	0	0.4	0.6	C	U
122	*Certhia familiaris**	0.1	0.3	0.4	C	N
123	*Oriolus oriolus*	0.8	0	0	R	D
124	*Lanius collurio*	0.1	0	0	O	N/A
125	*Lanius excubitor*	0	0.1	0.2	O	N
126	*Garrulus glandarius**	0.1	0.3	1.0	C	U
127	*Pica pica**	0.5	0.6	1.0	C	U
128	*Nucifraga caryocatactes*	0	0.1	0	O	N/A
129	*Corvus cornix**	1.1	1.0	1.0	C	N
130	*Corvus corax**	0.6	0.6	0.8	C	N
131	*Passer montanus**	0	0	0.1	O	N/A
132	*Fringilla coelebs**	1.0	1.0	1.0	C	N
133	*Fringilla montifringilla**	0.3	0.3	0.3	C	N
134	*Carduelis chloris**	0.1	1.0	1.0	C	U
135	*Carduelis spinus**	0.6	0.6	0.9	C	N
136	*Carduelis flammea**	0.3	0.4	0.5	C	N
137	*Loxia curvirostra**	0.3	0.2	0.3	C	N
138	*Loxia pytyopsittacus*	0.1	0	0.2	R	N
139	*Carpodacus erythrinus*	1.4	0.4	0	C	D
140	*Pinicola enucleator*	0	0	0.1	O	N/A

Number	Species	1989–2000	2001–2013	2014–2023	Status/ Abundance	Trend
141	*Pyrrhula pyrrhula**	0.8	0.9	1.1	C	N
142	*Coccothraustes coccothraustes**	0	0	0.2	R	N
143	*Calcarius lapponicus*	0.1	0	0	O	N/A
144	*Plectrophenax nivalis*	0	0.1	0.1	O	N
145	*Emberiza citrinella**	0.3	0.6	0.1	C	D
	Totals	108	102	114	69C; 29R; 47O	20 U; 18D; 78 N 29 N/A

Observations presented in three periods and recorded years of each species are divided with the same period years of Chaffinch Fringilla coelebs, *Finland's most common bird species [9]. C = Common (7 or more observation years); R = Rare (4–6 observation years) and O = Occasional (1–3 observation years); U = upwards (at least 0.4 up), D = downwards (at least 0.4 down) and N = no major trend up or down (less than 0.4 up or down), N/A = seen only once or during one year, so no population trend estimation possible.* **Species are seen to visit the bird feeding sites.*

Table 1.
Relative abundance of all bird species seen in Tervasaari, Savonranta, Finland from May 1989 to June 2023.

year, new nest boxes, of different sizes and shapes, have been built so that the final number of nest boxes is well over 50.

3. Results

Richness and relative abundance of all recorded bird species were studied from May 1989 until July 2023. The observations were grouped into three parts, the first part from 1989 to 2000, the second from 2001 to 2013 and the third from 2014 to 2023 (**Table 1**). A total of 145 bird species were identified over the years, 108 during the first period, 102 during the second and 114 during the third period. Abundance classification listed 69 (48%) species as common, 29 (20%) rare and 47 (32%) as occasional. The population trend was possible to identify only in species seen more than once and in more than one year. Of total species, 116 (80%) were such and in 20 (17%) species trend is going upwards, in 18 (16%) species trend is downwards and the remaining 78 (67%) species are not showing any clear changes. Bird feeding attracted 40 (28%) of species mainly during the winter but later also almost all year around as shown below.

4. Comments on some species

4.1 Black-throated Diver *Gavia arctica*

Black-throated Diver has been going upwards in Tervasaari waters same time when Great Crested Grebe *Podiceps cristatus* has gone down although not known if any linkage between these two changes exists.

4.2 Arctic geese

We did not see Arctic geese migration in the early years as we were using the house mainly during the summer. It is also, however, obvious that the amount of Arctic geese, especially the Barnacle Goose *Branta leucopsis* has had a heavy population increase in all of its northern distribution areas [10]. Winter survival of Barnacle Geese has increased enormously because of their access to higher-quality food in agricultural areas [11]. There are also indications that the migration route of the Arctic geese has moved more towards the west bringing the geese just over Tervasaari [12].

4.3 Birds of Prey and Owls

Predation is often density-dependent, and a higher population density of prey may feedback on the density of predator populations [2]. This was seen clearly in the numbers of Sparrow Hawk *Accipiter nisus*, Ural Owl *Strix uralensis* and Pygmy Owl *Glaucidium passerinum,* which have learned to visit our bird feeding sites. When Sparrow Hawk comes visible, all birds feeding in the area will 'freeze' to death, not moving even their eyes. The Red Squirrel *Sciurus vulgaris* regularly visits the feeding place and also is freezing when noting the hawk. Magpie *Pica pica* is the largest prey Sparrow Hawk that has been taken, so Squirrel has all reasons to be afraid. Pygmy and Ural Owl may come near the feeding places as there are many voles under the feeders but at least the first one takes readily also small Passerines forming a very important part (up to 40%) of its diet [13].

4.4 Mammal visitors

Feeding place food may bring many mammal predators to the site. Eurasian lynx *Lynx lynx*, Red fox *Vulpes vulpes* and Pine Marten *Martes martes* have learned to steal our fat seed bars and lard, especially if we are not present in the house. Also, Eurasian ermine *Mustela erminae* has been seen climbing the tree stump to reach the lard at a height of two meters. Raccoon Dog *Nyctereutes procyonoides* visit the feeding sites but get only some fallen sunflower seeds. Footprints of Wolverine *Gulo gulo* and Brown Bear *Ursus arctos* have been seen in our yard but not near the bird feeding tables. Sometimes Roe Deer *Capriolus capriolus* and White-tailed Deer *Odocoileus virginianus* may also feed some sunflower seeds on the ground.

4.5 Bullying

Some species, like Greenfinch *Carduelis chloris,* are born to be bullies towards all smaller species and often even towards the bigger ones, for example, against Bullfinch *Pyrrhula pyrrhula,* which are less aggressive. However, Greenfinch did not dare to disturb the feeding of Hawfinch *Coccothraustes coccothraustes* although harassing same time the Bulffinch. Hawfinch was not minding the presence of other birds on the feeding table. Siskin *Carduelis spinus* was often seen feeding side by side with Hawfinch despite its huge bill and bull neck. Great Tit *Parus major*, especially male birds are often bullying the smaller tits, like Eurasian Blue Tits *Cyanistes caeruleus*. When food is plenty, fighting for it is less intense – even bullying.

The Red Squirrel or Jay *Garrulus glandarius* often keep the feeding table occupied to those other birds or other Squirrels may not enter to feed themselves. However, agile Siskins and Willow Tits *Poecile montanus* often pick seeds despite the presence of large feeders.

4.6 Learning

Many species like Waxwings *Bombucilla garrulus* and Redpolls *Acanthis flammea* have learned to visit below the feeding table to eat the sunflower seeds, which woodpeckers and jays were dropping down. Robin *Erithacus rubecula* did that same but soon learned to go up to the feeding table. Some Blackbirds *Turdus merula* learned obviously by observing the woodpeckers how to eat directly fat seed bars, although their legs are not so good for hanging from the net protecting the bar. In addition, Siskins and Long-tailed Tits *Aegithalos caudatus* have been feeding on the fat seed bars without any problems.

4.7 Mutualism

Clear mutualism is when tits and blackbirds are waiting on the ground below the fat seed bars that woodpeckers will drop small parts for them as well.

4.8 Fight for the nest boxes

Great Tit and Eurasian Blue Tit are reducing the nest box competition by reserving their favourite boxes early in the winter. The problem comes later when the Pied Flycatcher *Ficedula hypoleuca* is returning from migration because it is a real fighter for the nest boxes. Male birds try to steal the box, especially from Eurasian Blue Tits. On 11/05/2021, three female Pied Flycatchers and one male kept the Eurasian Blue Tit female inside the nest box by force so that the male Blue Tit was not able to feed the incubating female. Later the Pied Flycatcher overtook that nestbox. When cleaning the boxes, there have been cases of the Pied Flycatcher having built its nest over the body of the Blue Tit. The Great Tit is stronger to defend its nesting and can sometimes even harm the Pied Flycatcher.

4.9 Fight for the twig nests

Eurasian Hobby *Falco subbuteo* was seen fighting with the Hooded Crow for a twig nest on 25/05/2017. Breeding of the Hooded Crow should have been over, so not known if it was just defending the same year-used nest.

4.10 Common Gull *Larus canus*

On 16/07/2020, when feeding the fish from our pier with breadcrumbs, three Common Gulls came to watch us (or rather small fish!). One Common Gull was much tamer than the others were and was less than 10 m from us. From the game camera photos, it was noted that one Common Gull was also visiting the feeding table.

4.11 Herring Gull *Larus argentatus*

One big-headed Herring Gull learned to visit the feeding table especially if some Rainbow Trout *Oncorhynchus mykiss* remains were available. The same Herring Gull was often following us during the fish feeding but not so close as the Common Gull above.

DOI: http://dx.doi.org/10.5772/intechopen.1002390

4.12 Wood Pigeon *Columba palumbus*

Wood Pigeons have also realised that below the bird feeding table, they get easy food. Two birds very often visited the site together but mostly eat only on the ground. At least one of them has learnt to get seeds also from the feeding table.

4.13 Grey-headed Woodpecker *Picus canus*

This large woodpecker is often very peaceful with other birds sharing the fat seed bars with tits without any problems. However, sometimes male Grey-headed Woodpeckers did not want to share the food with White-backed Woodpecker *Dendrocopos leucotos* male—both spreading their wings to drive the other away.

4.14 Great spotted Woodpecker *Dendrocopos major*

This species is a nuisance for smaller birds breeding in the nest boxes by breaking through the walls to reach the young ones. It has been noted that if feeding birds also during the early summer then the Woodpecker damages are much less. They bring their own young ones to feed the seeds in the feeding place instead of breaking the nest boxes. It seems always to be the male, which brings the young Woodpeckers to learn to eat sunflower seeds. First, the father opens the seeds for the young ones until they learn to do it themselves. The seed is placed into a small wood hole, which holds it tight until the hardcover is removed.

One fat female Great Spotted Woodpecker decided that she is the only bird who can eat sunflower seeds from our feeder. She attacked all other species from the size of Siskin to the size of Hooded Crow.

4.15 White-backed Woodpecker

Aggressive behaviour was noted between the male White-backed Woodpecker and the female Great Spotted Woodpecker very often. On 26/03/2016, three White-backed Woodpeckers tried to stop one Great Spotted Woodpecker to enter the feeding table.

4.16 Jay

Normally there were four up to seven Jays in the feeding but on 22/04/2019 there were 15 Jays in our front yard, obviously indicating some kind of return migration from their wintering area if not just presenting several families, which have been nesting nearby. On 16/04/2022, there were 5 + 5 (two families with the young ones) Jays fighting like cats and dogs, which family should be able to feed themselves. After a while, another family gave up and only one of the five groups remained in the feeding.

4.17 Magpie *Pica pica*

Feeding visits often occurred only very early in the morning and normally alone or not more than two together. One time, end of July 2019, ten Magpies were on the feeding site, which may again is a sign of some invasion. Magpies are learning to spy on the Jays when they are hiding the food from the feeding place. Not known how often Jays are returning in vain to these food caches, which Magpies have stolen.

4.18 Hooded Crow

In Tervasaari, Hooded Crow is a partial migrant and leaves the area during the winter. Autumn departure timing is not so easy to note but happens usually between the end of September and early November. After returning, the birds usually come to announce their presence, so the timing was also easier to record. Over the years, Hooded Crow returned to Tervasaari between 28 February and 29 March as shown in **Table 2**.

In 2014, we arrived from the UK too late in the spring to record the first arrival; also in 2017, the recorded time is far too late to present the first arrival. In folklore, we have a saying that the crows are returning on 24/2 [14] indicating that springtime starts now later than before 1950 despite the overall climate warming.

4.19 Raven *Corvus corax*

Ravens hold territory during the entire year near their nesting locality and will occasionally go for a 'winter feeding trip', for example, in rubbish dumps. Therefore, it is possible that the closing of rubbish dumps influences the winter distribution of the Ravens [15]. In old times, we had a wooden closure without cover for the household food waste. Then the Ravens brought even the young ones to feed next to the house, now the waste is in a closed composter and Ravens are not so regular visitors anymore. The nesting site was also affected by forest harvesting.

4.20 Eurasian Blue Tit

In 2017, Great Spotted Woodpecker ate the Eurasian Blue Tit young ones from a nest box in May. In early June, the same pair started a new nest in another nest box, some 50 m from the first box. Again, the Great Spotted Woodpecker broke the box wall and ate the young ones despite the plywood strengthening the wall of that box. The third time that pair nested in a third box behind our sauna, and finally succeeded. The young ones left that nest box on 24/07/2017 without any disturbance from the Great Spotted Woodpecker. Unfortunately, summer feeding started only regularly after 2017 as that stopped almost completely these Great Spotted Woodpecker attacks on smaller bird nest boxes.

4.21 Chaffinch

During the spring migration, there has been some 50 Chaffinch in front of the house but only three or four of those will dare to enter the feeding table. Those pairs that remain in the site for breeding are later more regular in the feeding and will also encourage the later returning Bramblings *Fringilla montifringilla* to enter the feeding table.

4.22 Summer feeding

If feeding was continued until the end of June, many species, like Bullfinch, Chaffinch, Great Tit, Greenfinch, Hooded Crow, Siskins and earlier mentioned Great Spotted Woodpecker, bring their young ones to the feeding table. It is almost always the male first feeding the young ones while teaching them how to eat sunflower seeds and groundnuts. Interestingly, Eurasian Blue Tit does not bring the young ones but visits the feeder and takes food to the young ones waiting nearby. Later well-flying young ones come to feeding table independently.

DOI: http://dx.doi.org/10.5772/intechopen.1002390

Year	2006	2008	2009	2010	2011	2012	2013	2015	2016	2017	2018	2019	2020	2021	2022	2023
Arrival	12/3	10/3	16/3	18/3	27/3	11/3	19/3	28/2	10/3	22/4	29/3	18/3	09/3	19/3	16/3	17/3

Table 2.
Spring time arrival of the Hooded Crows Corvus corone cornix on the Tervasaari from 2006 to 2023.

4.23 Forest harvesting

Our neighbours started to harvest their forests in 2019 and 2023 less than 100 and 500 meters away from our forest. The disappearance of old spruce forests affected immediately the abundance of such species as Willow Tit, Crested Tit *Lophophanes cristatus* and maybe also Coal Tit *Periparus ater.* Next summer will show if these species have disappeared forever as it now started to look during the 2023 winter feeding. Elsewhere in Finland modification, fragmentation and loss of boreal forest are known to have a major role for instance in the decline of the Crested and Willow Tit populations [16, 17].

5. Discussion

In this study, we noted an increase in Black-throated Loon and a decrease in Great Crested Grebe. That same change was noted in a country-wide inland water study [18] in which 10 of the 16 common waterfowl species showed significant declines and only three species (Whooper Swan *Cygnus cygnus*, Mallard *Anas platyrhynchos* and already mentioned Black-throated Loon), have increased in 1986–2018.

Our results show a decline in the populations of the Little Gull *Hydrocoloeus minutus* and Black-headed Gull *Larus ridibundus*. The overall decrease in waterfowl has been connected to the low number of Black-headed Gulls [18]. While breeding in the large gull colonies, the waterfowl enjoyed good protection against raptor and mammal predation [19].

On the European scale [20], the status of the Common Snipe *Gallinago gallinago* has changed from least concern to vulnerable primarily due to habitat loss and degradation in its breeding grounds. Our results confirm this after the neighbour was drying the nearby marsh where this bird was breeding. Common Snipe has not been recorded in this study since 2011. Marsh drying was destroying also the breeding place of Little Gulls and Black-headed Gulls.

A different story is the Great Spotted Woodpecker, which on a European scale [20] has a declining population due to unsustainable forestry practices but in our area is doing very well mainly due to feeding operations. Rapid increases of Grey-headed Woodpeckers and Blue Tits have been noted in this study as well as in the country-wide study [16] and the most likely reason is again the improved food supply due to winter-feeding.

Wintering bird populations can track habitat and climate change better than breeding populations [15]. A typical example is the Eurasian Jay, which has benefitted from global warming since its population range has expanded in Finland [21]. The Eurasian Jay increased in our study period from 0.1 to 1.0 and is also more common and abundant elsewhere in the Finnish winter-feeding sites [22].

6. Conclusions

The range and distribution of bird species worldwide are shifting in response to climate change, which can affect species occupancy either directly by exceeding physiological constraints (e.g., heat tolerance) or indirectly by altering resource availability or other ecological conditions [23]. Since 1980, a loss of 560–620 million individual birds has been reported, that is, 17–19% of the overall breeding bird abundance in the area of the European Union [24, 25]. Based on regional and global alarming reports on climate change, one would have expected much more drastic changes in our study area. A total of 145 bird species were identified during 34 years, which represents 32% of the European breeding species [25]. Only 16% of the 116 regular species have been recorded to go downwards while almost the same amount (17%) of the species have shown an upward trend in their occurrence between 1989 and 2023. Naturally, the small study site is a well-protected area from many anthropogenic impacts, including water and air pollution, urban noise and light disturbance, and the feeding of the birds and numerous nest boxes have assisted their survival and diversity. The future of Tervasaari birdlife does not look as good as two next-door neighbours were harvesting heavily their forests in 2019 and 2023. Remains to be seen how badly the disappearance of the old spruce forest will affect the abundance of birds listed in this study.

Acknowledgements

The authors want to express their best thanks to the Finnish photographers, namely Esko Rajala, Jari Peltomäki and Ari Rantamäki, who kindly forwarded their amazing images for this chapter.

Author details

Heimo Mikkola[1*] and Anita Mikkola[2]

1 University of Eastern Finland, Kuopio and Joensuu, Finland

2 University of Oulu, Finland

*Address all correspondence to: heimomikkola@aol.com

References

[1] Berthold P. Patterns of avian migration in light of current 'greenhouse' effects. A central European perspective. In: Acta XX Congressus Internationalis Ornithologici, Christchurch, 2-9 December 1990. Wellington, N.Z.: Ornithological Congress Trust Board; 1991. pp. 780-786. DOI: 10.3962/bhl.title.143160

[2] Møller AP, Berthold P, Fiedler W. The challenge of future research on climate change and avian biology. Advances in Ecological Research. 2004;**35**:237-245. DOI: 10.1016/S0065-2504(04)35011-7

[3] Li X, Liu Y, Zhu Y. The effects of climate change on birds and approaches to response. IOP Conference Series, Earth and Environmental Science. 2022;**1011**:012054. DOI: 10.1088/1755-1315/1011/1/012054

[4] Brambilla M, Scridel D, Bazzi G, Ilahlane L, Iemma A, Pedrini P, et al. Species interactions and climate change: How the disruption of species co-occurrence will impact on an avian forest guild. Global Change Biology. 2020;**26**:1212-1224. DOI: 10.1111/gcb.14953

[5] Schofield LN, Siegel RB, Loffland HL. Modelling climate-driven range shifts in population of two bird species limited by habitat independent of climate. Ecosphere. 2023;**14**:e4408. DOI: 10.1002/ecs2.4408

[6] BirdLife International. State of the World's Birds 2022: Insights and Solutions for the Biodiversity Crisis. Cambridge, UK; 2020. p. 45. Available from: https://www.birdlife.org>state-of-the-world-birds-2022/ [Accessed: February 10, 2023]

[7] European Environment Agency (EEA). Bird Populations: Latest Status and Trends. Available from: https://www.eea.europa.eu/themes/biodiversity/state-of-nature-in-the-eu/bird-populations-what-are-the [Accessed: February 20, 2023]

[8] EEA. State of Nature in the EU. Vol. 10. Copenhagen; 2020. pp. 1-146. Available from: https://www.eea.europa.eu>state-of-in-europa-a-healthcheck/bird-populations-lateststarus-and-trends [Accessed: February 10, 2023]

[9] LuontoPortti. Ennätyslajeja:Suomen yleisimmät linnut. 2021. Available from: https://luontoportti.com/577/peippo [Accessed: February 21, 2023]

[10] Mikkola H. Management of the Barnacle Goose (*Branta leucopsis*) in Finland: Conservation versus Hunting. Chapter 8. In: Mikkola H, editor. Birds—Challenges and Opportunities for Business, Conservation and Research. London, UK, London: InTech; 2021. pp. 127-142. DOI: 10.5772/intechopen. 82911

[11] Lameris TK, van der Jeugd HP, Eichorn G, Dokter AM, Bouten W, Boom MP, et al. Arctic geese tune migration to a warming climate but still suffer from a phenological mismatch. Current Biology. 2018;**28**(15):2467-2473.e4. DOI: 10.1016/j.cub.2018.05.077/

[12] Hoag H. Climate Change has Barnacle Geese Racing North. Arctic Today. Rogoff's Polar Connection; 2018. p. 6. Available from: alice@arctictoday.com

[13] Mikkola H. Owls of Europe. Calton: T & A D Poyser; 1983. p. 397

[14] Vilkuna K. Vuotuinen ajantieto. Helsinki: Otava; 1950. p. 364

[15] Jokimäki J, Kaisanlahti-Jokimäki M-L, Suhonen J. Long-term winter

population trends of corvids in relation to urbanization; and climate at northern latitudes. Animals. 1820;**2022**:12. DOI: 10.3390/ani 12141820

[16] Lehikoinen A, Tirri I. Talvilintujen alueelliset runsausmuutokset (summary: Changes in regional wintering abundances of birds in Finland). Linnut-Vuosikirja. 2020;**2020**:18-29

[17] Kumpula S, Vatka E, Orell M, Rytkönen S. Effects of forest management on the spatial distribution of the willow tit (*Poecile montanus*). Forest Ecology and Management. 2023;**529**:1-11. DOI: 10.1016/j.foreco.2022.120694

[18] Laaksonen T, Lehikoinen A, Pöysä H, Sirkiä P, Ikonen K. Sisävesien vesilintujen kannanvaihtelut 1986-2018 (Summary: Inland waterfowl population trends 1986-2018). Linnut-Vuosikirja. 2019;**2019**:46-55

[19] Pöysä H, Lammi E, Pöysä S, Väänänen V-M. Collapse of a protector species drives secondary endangerment in waterbird communities. Biological Conservation. 2019;**230**:75-81

[20] BirdLife International. New BirdLife Report: 1 in 5 Bird Species in Europe is Threatened by Extinction. Brussels. 2021. Available from: https://www.birdlife.org/news/2021/10/14/press-release [Accessed: February 10, 2023]

[21] Valkama J, Vepsäläinen V, Lehikoinen A. The Third Breeding Bird Atlas of Finland; Luonnontieteellinen keskusmuseo ja ympäristöministeriö. 2011. Available from: http://atlas3.lintuatlas.fi [Accessed: May 4, 2023] (In Finnish with an English summary)

[22] Väisänen RA. Frequency and abundance of 61 bird species at feeding sites in Finland in 1990/1991-2019/2020. Linnut-Vuosikirja. 2021;**2020**:30-45 (In Finnish with an English summary)

[23] Virkkala R, Marmion M, Heikkinen RK, Thuiller W, Luoto M. Predicting range shifts of northern bird species: Influence of modelling technique and topography. Acta Oecologica. 2010;**36**:269-281

[24] McLean N, Krunk LEB, van der Jeugd HP, van de Pol M. Warming temperatures drive at least half of the magnitude of long-term trait changes in European birds. PNAS. 2022;**119**:10. DOI: 10.1073/pnas.21054/6119

[25] Burns F, Eaton MA, Burfield IJ, Klvaňova A, Šilarová E, Staneva A, et al. Abundance decline in the avifauna of the European Union reveals cross-continental similarities in birddiversity change. Ecology and Evolution. 2021;**11**(23):16647-16660. DOI: 10.1002/ece3.8282

Chapter 6

Influence of Agriculture on the Resilience and Trophic Cascade in a Savanna Bird Community of the Serengeti Ecosystem, Tanzania

Ally K. Nkwabi and Pius Y. Kavana

Abstract

Agriculture is one of the factors contributing to the global depletion of natural resources. A total of 228 bird species were recorded in the savanna of Serengeti National Park and 146 bird species in agricultural areas. The richness of bird species endemic to the Serengeti ecosystem, regional and Africa-wide geographical ranges were all higher inside Serengeti National Park compared to agricultural land. However, a greater proportion of endemic bird species dropped out in agriculture compared to the proportions of species with wider geographic ranges. With respect to the trophic level in agriculture, there was a greater proportional decrease of species (compared to savanna) at the predator trophic level (vertebrate feeders) than at the insectivore and herbivore trophic levels. Therefore, higher feeding levels were less able to tolerate changes in habitat structure due to agriculture than species at lower feeding levels. These results suggest that Park managers should encourage through education and participation of local communities on the importance of improving their agricultural habitats by promoting the regeneration of patches of native vegetation so that more vulnerable species, such as endemics, and raptors can exist there.

Keywords: conservation, climate change, birds' distribution, disturbance, environmental drivers, human modified habitats

1. Introduction

Agriculture is one of the factors that contribute to the global depletion of natural resources [1–4]. Land clearing and encroachment are the main contributors to the decline of wildlife and local species extinction worldwide [5–8]. The increase in human population has resulted in the rise in demand for land for settlement and food to meet the body requirements of people. Over the last four decades, areas adjacent to Serengeti National Park have experienced huge population growth [9, 10]. The period between 1957 and 1977 recorded the highest rate of population increase, some 10% per annum. Of this, only 3.4% was due to natural increase while the rest was due to immigration [9–11]. The current population in the seven wards to the west of the park is over two million with an annual growth rate exceeding the national average of 2.9% [11, 12].

Human activities such as overgrazing, deforestation, bushfires, mining, urbanization, and cultivation as well as climate change have aggravated the ability of bird species to disperse into new habitats [13, 14]. These two factors, therefore, have resulted in degradation or modification of habitat structure with consequent decline of avian food resources.

Tanzania has reserved about 40% of its land as protected areas [15]; these are distributed throughout the country and are largely located adjacent to rural communities. The majority of these rural communities depend on crops and livestock for their livelihoods; thus, a considerable number of people have settled around buffer zones surrounding protected areas where they believe that the land is fertile and there is sufficient grazing land [9, 16, 17]. Kideghesho et al. [11] showed that there was a 52% increase in livestock from 175,680 to 266,624 from 1990 to 2011, respectively within areas adjacent to the western Serengeti National Park. At the same time, the expansion of arable land and settlement in villages has led to the shrinkage of grazing land for livestock [9, 10]. This density is now well above the livestock carrying capacity, which was already exceeded a decade ago [11, 18]. Livestock numbers have been reported to increase at the peripheral of protected areas likely to impact severely on the native vegetation and expected to influence wildlife daily activities in the park.

1.1 Interaction of environmental change with agriculture areas

In tropical regions, the main climate variable is rainfall rather than the temperature which is the dominant variable at higher latitudes. Rain in the Serengeti ecosystem typically falls in a bimodal pattern, with the long rains during March–May and the short rains during November–December [19]. However, the rains can fuse into one long period, particularly in the north, or the short rains can fail entirely, especially in the southeast [19]. Most importantly the Serengeti ecosystem has a strong gradient in the total annual rainfall from 500 mm in the southeast to 1100 mm in the north and west [19]. The wettest part of the ecosystem is in the agricultural region of northwest Serengeti National Park around Mugumu and Tarime [20, 21]. If hypothetically the Serengeti National Park becomes drier than the savanna bird community within, it would move up the rainfall gradient into the agricultural region. Although most bird species avoid agricultural areas, nearly 30% of all birds regularly or occasionally use such habitats [22, 23], often providing important ecosystem services, such as pest control, pollination, seed dispersal, and nutrient deposition [22, 24].

The main environmental driver in Serengeti National Park is rainfall, with temperature being of lesser importance, although it may still have a subsidiary influence [25, 26]. Animal communities in Serengeti National Park respond to changes in rainfall; if conditions become drier communities will move toward wetter regions. To understand how climate change might affect the Serengeti National Park through changes in rainfall the bird community was taken as a case study. Within the Serengeti National Park, birds are constrained in how far they can move along the rainfall gradient in response to climate change because there are barriers from the human settlement at the edge of the park boundary in the north and west. If conditions become drier, birds will move toward wetter areas which now are severely modified by agriculture. Trees are largely absent, native grasses are replaced by monocultures of domestic crops, and shrubs become abundant along hedgerows. As a result bird species have responded differently to this habitat modification in the village lands [1, 23, 27, 28].

Climate change has already affected bird distributions in the northern temperate regions such as Europe and North America [29–33]. Far less is known about how climate can affect the distribution of birds in the tropics. It has been postulated that savanna woodlands make up most of Africa's tropical tree cover, but savannas are widely thought to be declining rapidly under both direct human impacts and indirect human-induced climate change [34–36]. With climate change, species' geographic distributions are expected to change. Bird species may become locally extinct in existing protected areas that are no longer environmentally suitable but cannot move into more suitable areas because of habitat modification by humans [36–38]. Therefore, the conservation of vulnerable bird species may require new areas that are currently both unprotected and climatically unsuitable but could become suitable later [39–43].

1.2 The importance of disturbance in ecological communities

The number of relatively undisturbed ecosystems in the world is decreasing rapidly and living organisms are confronted with man-made challenges, including climate change and habitat destruction. By the year 2050, 15–37% of existing animal and plant species on earth are predicted to become extinct [37, 44, 45] and half of all species on earth may experience extinction by 2100 [46]. It is important to predict how current man-made challenges will affect key species in ecosystems, so that, if possible, steps can be taken to ameliorate these threats.

There is much scientific concern that the loss of living organisms will reduce the capacity of our ecosystems to provide important services for human populations. Examples of such services are primary production, nutrient cycling, mineral cycling, and pollution control [47–50]. In addition, the consequences of species extinction on ecosystems may not be detected until it is too late to reverse the loss [51]. Some ecosystems respond to disturbances by losing a few species at first but many more later on; so a need for conservation authorities management to reverse species loss would benefit from early detection [51, 52]. Responses of biota including changes in physiology, productivity, and growth have been caused by human activities and climate change acting upon their habitat [53, 54], and changes in species distribution and abundance due to migration and range shifts have also been caused by habitat loss [55].

Tolerance limits are usually defined in terms of environmental parameters such as temperature, rainfall, altitude, or habitats but they can be complex involving a mixture of all these plus tolerance of predation [56]. Tolerance limits are situations in which living organisms cannot distribute, survive, grow, or reproduce under limited environmental factors [56]. Distribution of organisms can be endemic (with narrow distribution), regional, or Africa-wide ranging depending on the environmental factors of a particular area.

There has been much study of the pattern of wide-ranging or narrow-ranging species in terms of wide tolerance or narrow tolerance of these factors, in order to understand what causes areas of high diversity in the world [57–59]. This study uses the variation in distribution and hence tolerance limits to predict which species are most susceptible to man-made disturbances in the form of agricultural modification of habitat.

Agriculture around protected areas is one of the major threats to wildlife conservation in Tanzania [1, 9, 60–62]. Cultivation and livestock husbandry around national parks and game reserves affect natural resource conservation in Tanzania [63, 64]. The confinement of high densities of livestock on village lands causes overgrazing, soil erosion, and siltation of water bodies ultimately causing ecological imbalance, which interferes with ecological resilience [9, 10, 22, 65, 66]. Due to inadequate and poor quality pasture on these village lands, the livestock owners in some villages of western Serengeti National Park have been pressing the government to legalize access to critical grazing and watering points in Grumeti, Ikorongo, and Kijereshi Game Reserves [11, 66]. These villagers, however, are continuing to use the areas illegally in order to survive. Little information is available to describe how birds with different tolerance limits or feeding levels are affected by these habitat disturbances from agricultural modification.

The main objective of the study was to examine the effects of agriculture on the trophic cascade in savanna bird communities. The study had two specific objectives: First, to compare how species of birds with narrow ecological tolerance limits respond to anthropogenic disturbance of habitat (agriculture) in comparison to those with wide tolerance limits (Africa-wide species). Second, to compare how savanna bird species at different trophic levels respond to disturbance in human-modified habitats such as agriculture.

We hypothesized that endemic bird species with narrow ecological tolerance limits lose a greater proportion of species through agricultural modification of habitats than do bird species with wide ecological tolerance limits. Also, bird species in higher feeding guilds lose a greater proportion of species through agricultural modification than do species at lower feeding guilds.

2. Materials and methods

2.1 Study area

The study was conducted in Serengeti National Park (14,763 km^2) located between 33°50′–35°20′ E and 1°28′–3°17′ S (**Figure 1**). The park occupies an extensive upland area varying in elevation from 1162 m at Speke's Gulf to 1860 m above sea level in the northeast [67]. Serengeti National Park is characterized by grassland, savanna, closed canopy riverine forest, rocky outcrops, rivers, and alkaline lakes (**Figure 1**).

The climate shows a relatively constant monthly mean and maximum temperature of 27–28°C taken daily in the morning (9:00 AM) and afternoon (3:00 PM) at Seronera [19]. The minimum temperature varies from 16°C in the hot months of October–March to 13°C during May–August. Rain typically falls in a bimodal pattern, with the long rains during March–May and the short rains in November–December [19]. However, the rains can fuse into one long period, particularly in the north, or the short rains can fail entirely, especially in the southeast of the park [67].

2.2 Bird survey in agricultural and savanna habitats

Data in Serengeti National Park and in agricultural areas immediately outside the Park were obtained by similar methods to allow comparisons of the bird communities in the two study areas. The bird community was sampled along road transects in native savanna and through villages. Both transect and point count methods were used to record bird species, their numbers, and their habitats. Transects in both savanna and in agriculture were established using roads. A total of five set points of 100 m × 100 m along each transect in savanna and agriculture was established, marked by using the Global Positioning System (GPS) Garmin 12 XL made in Taiwan. Points along transects were spaced 2 km apart.

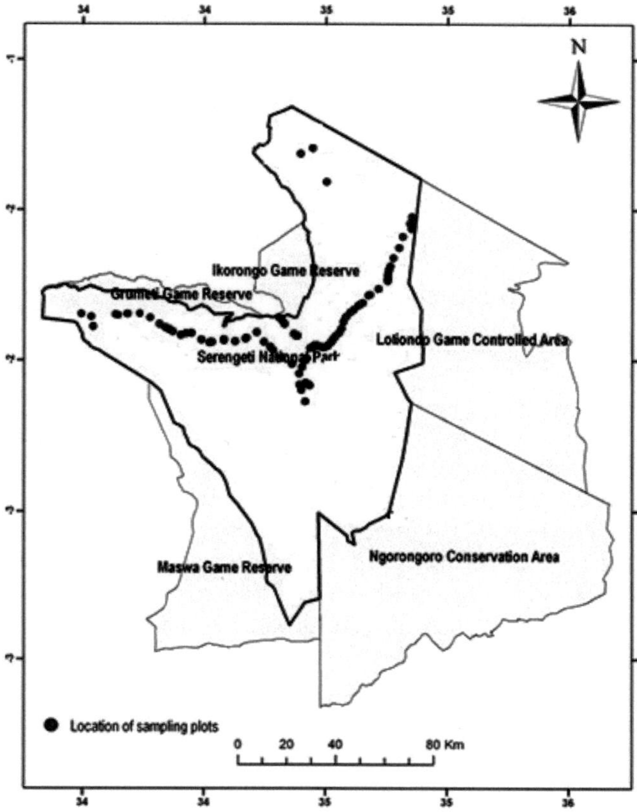

Figure 1.
Map of Serengeti National Park showing vegetation types. Black dots show bird sampling plots (Map provided by the Tanzania Wildlife Research Institute-GIS, 2023).

At the five set points established along each transect birds were recorded by sight or by hearing the call for 10 minutes. Transects were driven in the morning from 0615 to 1100 hours. The vehicle traveled at 3 km per hour and observers on each side of the vehicle recorded birds within 50 m on either side of the vehicle. For the agricultural area, only data from those parts of transects that fell within cultivation or habitations were used in this analysis. Cultivation comprised small fields of crops, hedgerows, and small areas used for grazing, usually 2–3 ha in size. Cultivation extended from 0.2 to 1 km on either side of the transects.

Three transects were established in agricultural areas. One was from Fort Ikoma through Mugumu town to Tabora B gate (80 km), a second ran from Mugumu Town to the Park boundary at Nyamburi village (40 km), and the third was from Fort Ikoma park headquarters to Isenye village (60 km). Transects were repeated every year between June 1997 and December 2011, the same observers who worked on the long-term project of the Serengeti Biodiversity Program conducted the surveys. Data from all transects were pooled across years to construct the sample for agricultural land.

Six transects within the Serengeti National Park provided the sample for the native savanna. These were: (i) Along the Mbalageti River (40 km), (ii) Along the southern edge of the woodlands where they meet the plains near Seronera (20 km), (iii) In the central woodlands near Banagi hill (20 km), (iv) In the northeast around Togoro and

Lobo (20 km), (v) In the northwest around Tabora (40 km), and (vi) In the far west around Ndabaka (40 km). These were a subset of a much larger sample of permanent road transects surveyed by the Serengeti Biodiversity Program. These transects were counted at intervals of six months from 1997 to December 2011. Data were summed across years for agricultural areas.

3. Data analysis

In comparing species richness between agricultural areas and the savanna of the Serengeti National Park the data were divided into two ways. First, species were examined according to differences in their geographical range within Africa as published in Sinclair and Ryan [68] and the seven volumes of the handbook "The Birds of Africa" [69, 70]. Three groups of geographical ranges were considered: (i) local endemic species found only in the Serengeti ecosystem, (ii) regional species confined to East Africa, and (iii) Africa-wide species (see Appendix 6).

Secondly, bird species were divided according to their feeding guilds, Simberloff and Dayan [71] defined a feeding guild as a group of animals that feeds on similar food types. A feeding guild can be a subset of a trophic level. Birds that feed on other birds, rodents, reptiles, amphibians, and fish were classified as vertebrate feeders (e.g. raptors, owls, and herons). Insectivorous birds were defined (for convenience) in this study as those that feed on arthropods, which includes insects, and also non-insect arthropods such as spiders and centipedes. Birds that feed on herbs, seeds, and fruits were all grouped in a feeding guild of herbivores (see Appendix 6).

Bird data were tested for normality using Kolmogorov-Smirnov (KS) test [72]. The data were not normally distributed even after transformation. Therefore a comparison of birds between agricultural land and the natural savanna of the Serengeti National Park was compared using the non-parametric test [72]. Bird species richness was estimated using the method of rarefaction given in the program of ecological methodology version 7.1 [73]. Because the sample within the protected area was much larger than that for agricultural areas the 'rarefaction' method was used to accommodate the difference in sample sizes. Rarefaction is a statistical method for estimating the species richness in a given sample of individuals. It allows two samples of different sizes, representing different sampling efforts, to be compared [74]. For example, within a given category of data, such as vertebrate feeders in agriculture, the program estimates the probable number of species in different sample sizes say 50, 100, 500 individuals etc. up to the total sample size. The same procedure was used for the sample from savanna with which the data for agriculture are being compared. For all tests $P < 0.05$ was considered significant. There are two restrictions on the use of this method [74]. First, the groups of species being compared must be from the same feeding guild in each habitat. Second, the sampling method for estimating abundance must be similar.

Shannon-Wiener diversity (H') indices were tested for significant differences by using the randomization test [75] to detect whether Savanna bird diversity changed when habitats were modified by agriculture [76]. Comparisons of abundance in agriculture and natural Savanna within each bird feeding guild used the Mann-Whitney U-test for independent samples [72].

4. Results

4.1 Agricultural effects on the resilience of the bird community

A total of 228 bird species were recorded in the savanna of Serengeti National Park and 146 bird species in agricultural areas. Estimates for species in the three range categories are presented in **Figure 2**. In agriculture, a sample of 3233 individuals was obtained which comprised of 88 species. In the savanna, a total of 150 species were recorded but for the same sample size as that for agriculture 97 species were estimated using Rarefaction. The rarefaction curve for Africa-wide birds showed higher species richness inside Serengeti National Park than in agricultural land. Initially, with the small sample size Africa-wide species in agriculture were greater in number than for the same sample size in the Serengeti National Park. However, as sample size increased the number of species in agriculture leveled out while those in Serengeti National Park continued to increase. Although there were more species in Serengeti National Park the difference was not great; some 9.3% of species were not observed in agriculture. The broad habitat niches of these species allowed them to tolerate the habitat modifications from agriculture.

For regional bird species in East Africa, a total of 55 species were sampled from the agricultural areas. The total number of bird species recorded in the Savanna was 109 but for the same sample as that for agriculture, the sample size in Savanna was reduced to 83 species. For local endemic bird species, 5 species were recorded in agricultural areas and 8 in Savanna for the same sample size but the total number in Savanna was 17. From rarefaction curves in **Figure 2**, the number of savanna bird species continued to increase as the sample size of individuals increased whereas the number of endemics in agriculture reached an asymptote. This result indicates that if a larger sample size had been obtained in agriculture there would be even larger

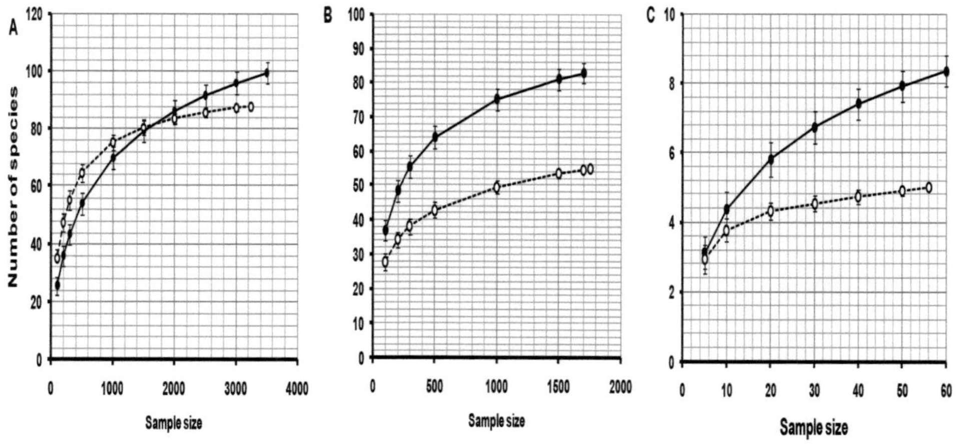

Figure 2.
Rarefaction curves for the number of bird species relative to the number of individuals counted along transects in agriculture (open circles) and the adjacent native savanna of Serengeti National park (closed circles) with respect to their geographical range in Africa. A. Africa-wide species. B. Regional species with an East African distribution. C. Local endemic species confined to the Serengeti ecosystem or areas close by. Vertical lines are standard error.

differences in the number of endemic species. The rarefaction curves show that the more restricted the geographical range of species the greater the loss of species when habitats were modified by agriculture. Thus, the proportion of species not observed in agriculturally modified habitats when compared to savanna was 9% for Africa-wide species, 34% for regional species, and 44% for local endemic species.

4.2 Effects of agriculture on the trophic cascade and food web stability

A comparison of bird diversity using the Shannon-Wiener index shows a difference between Serengeti National Park and agricultural areas. Herbivores ($H' = 3.261$) and insectivores ($H' = 3.432$) in agricultural areas have higher species diversity when compared with herbivores ($H' = 2.897$) and insectivores ($H' = 3.359$) in natural Savanna of Serengeti National Park. In contrast, birds that feed on vertebrates in natural savanna have higher species diversity ($H' = 2.620$) than in agricultural areas ($H' = 2.236$; **Figure 3**). These differences in species diversity between Serengeti National Park habitats and agricultural areas in the three comparisons above were significant ($P < 0.05$).

Rarefaction estimates for species in the three feeding guilds are presented in **Figure 4**. For herbivores, a total of 51 species were recorded from a sample of 2350. For the same sample size in Savanna, there were 60 species estimated but the total was 79 species in all from the total sample. Insectivores are the dominant guild in the avifauna and so more species were recorded for this group than for herbivores or

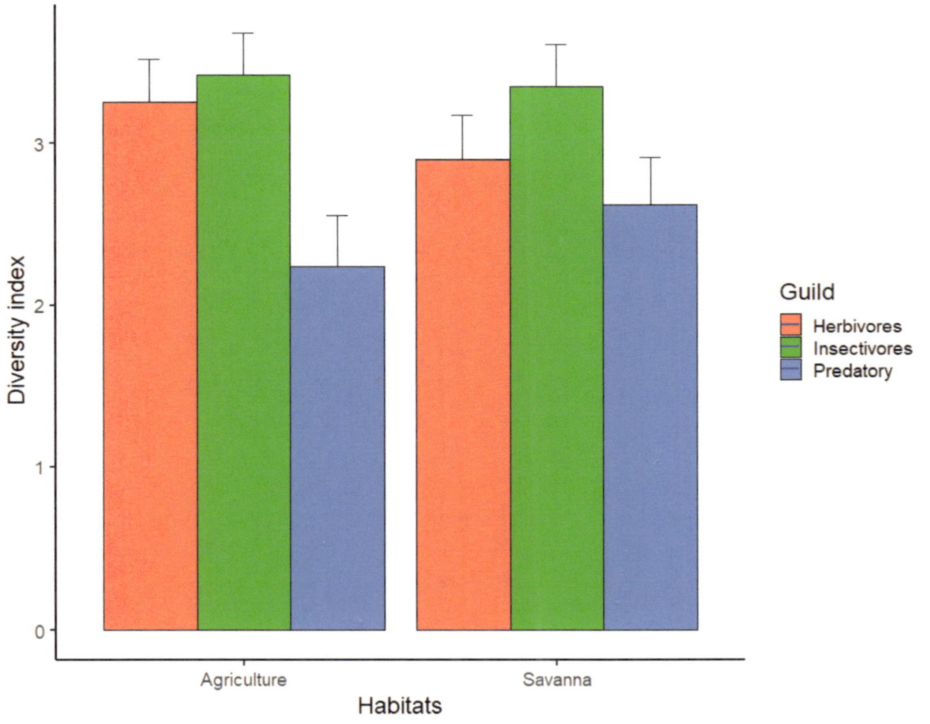

Figure 3.
Diversity indices for bird species observed and categorized according to their feeding guilds along transects in agricultural land and the adjacent native Savanna of Serengeti National Park. Vertical lines represent standard errors.

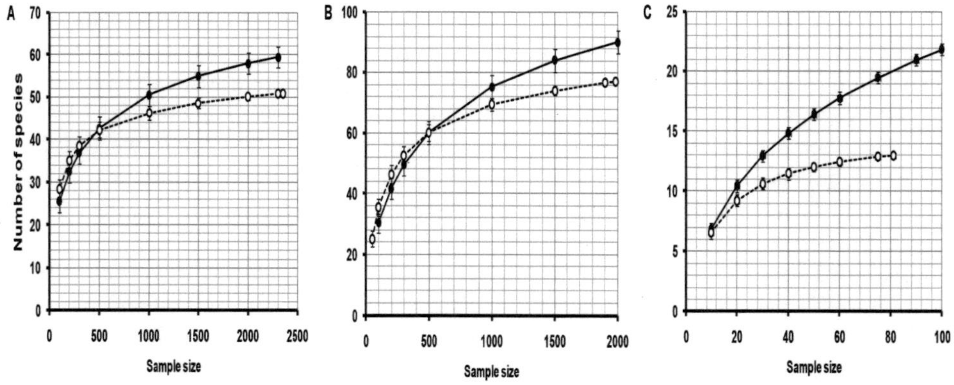

Figure 4.
Rarefaction curves for the number of bird species relative to the number of individuals counted along transects in agriculture (open circles) and the adjacent native savanna of Serengeti National Park (closed circles) with respect to their feeding guild. A. Herbivore species. B. Insectivore species. C. Birds that feed on vertebrates. Vertical lines represent standard errors.

vertebrate feeders. In agriculture, the total sample of 1983 produced 77 insectivores observed species. For the same sample size in the savanna, rarefaction estimated 90 species although the total number of species was 145 for the complete sample. There were 13 species of vertebrate feeders from a sample of 81 in agriculture. The equivalent number of species in the savanna was 21 from the same sample size with a total of 47 species recorded from the complete savanna sample. In summary, agriculture had 15% fewer herbivore species, 14% fewer insectivore species, and 38% fewer vertebrate feeders than those in savannah (**Figure 4**).

Differences in the abundances of birds in agriculture ($n_{agriculture}$) and Savanna ($n_{savanna}$) were tested with the Mann-Whitney U test. At all three trophic levels, there was a significantly higher abundance of birds in the Savanna than in agriculture. Thus, for herbivores $U = 763.5$, ($n_{savanna} = 70$, $n_{agriculture} = 43$, $P < 0.00001$), for insectivores $U = 1862.0$, ($n_{savanna} = 109$, $n_{agriculture} = 52$, $P < 0.0004$) and for vertebrate feeder $U = 85.0$, ($n_{savanna} = 23$, $n_{agriculture} = 14$, $P < 0.016$).

5. Discussion

This chapter examines the effects of disturbance on the distribution of bird communities in the savanna, and in particular the effects of habitat modification through agriculture. In the introduction, it was predicted that if wide-ranging species had broad habitat tolerance then such species should be able to tolerate agricultural modifications better than bird species with small ranges and narrow habitat requirements. Secondly, it was predicted that since the effects of disturbance are magnified up the trophic food chain [77] agriculture should have greater effects through the loss of species on higher trophic levels. These two hypotheses are discussed in detail below.

5.1 Agricultural effects on the resilience of the bird communities

These results indicate that birds in agricultural land may be affected by human activities through the planting of crops, the change in the herbaceous and shrub communities, and the removal of trees, all of which birds use for foraging, nesting, and perching. In general, the evidence supports the hypothesis that major changes to habitats such as agricultural modification causes bird communities to change, with species with broad niches and wide tolerance limits being more able to withstand the change than those with narrow ecological niches. Using the range of a species in Africa as the indicator of wide or narrow tolerance limits it was predicted that a greater proportion of local endemics would be lost in agricultural habitat than would species with broader habitat preferences, the Africa-wide species. The results are consistent with this hypothesis.

Local bird species with narrow geographical ranges have a lower tolerance to habitat change [78] and therefore are less able to survive in agricultural land compared to Africa-wide species with wide geographical ranges and ecological tolerance. The greater decrease of local endemics bird species when habitats changed from natural Savanna to agriculture was probably due to a combination of changes in habitat structure in agricultural land, such as loss of trees, monocultures of crops, and perhaps more exotic plants which do not provide suitable habitat for their survival. Habitat loss, overhunting, and animal disease could be factors contributing to the greater loss of local endemic birds in agricultural land. These factors promote protected area isolation by restricting birds' movement and dispersal.

These findings are consistent with other studies. Waltert et al. [79] reported a decline of endemic birds; they observed species composition gradually changed as the habitat transformed from natural forest to agricultural land in Sulawesi, Indonesia. Bolwig et al. [80] recorded biodiversity change in the forest under agricultural intensification in Uganda's farmed landscapes and reported a decrease in bird abundance and species richness with land use intensification. Neuschulz et al. [81] showed that patches of natural forest in South Africa were essential for the conservation of local endemic bird species while generalist species were able to expand into and tolerate modified and degraded forest habitat.

Agricultural activities outside protected areas have resulted in changes in vegetation from diverse natural plant assemblages to cultivated land with large areas of single domestic crops and low diversity of indigenous plant species; crops also result in an increase in exotic plant species that are not suitable for the survival of local endemic bird species [82, 83]. Many studies of birds that compared protected areas with agricultural land have shown a dramatic decrease in bird species in the agricultural area [1, 20, 84–87]. Similarly, Moreau [88] reported that intense habitat alteration by human activities has contributed to the disappearance of many species of birds. Land-use changes have also led to land degradation, which in turn negatively affects the species richness of birds [89–95].

5.2 Agricultural effects on the trophic cascade and food web stability

It was predicted that disturbance was intensified up the food chain so that more species would be lost at higher trophic levels due to agriculture. There was a drop in vertebrate feeders when compared to the lower trophic levels. The sharp decrease in the number of vertebrate feeders in agricultural land may have resulted from an increase in human activities through the removal of trees that birds use for nesting and perching. Another possibility is that the food items present in agricultural land may differ in type, abundance, and availability. Because disturbance becomes magnified up the trophic level, vertebrate feeder experiences greater effects than species at lower trophic levels and the results of this study have indicated this effect. This change in the trophic structure could result in more marked fluctuations in the biota if there is a top-down control in the food chain of the Serengeti ecosystem. Top-down control is defined as the limitation of population numbers at one level by species at higher trophic levels [96, 97]. There is considerable evidence that top-down control commonly occurs in all terrestrial ecosystems, as synthesized in Terborgh and Estes [98] and Estes et al. [99]. Top-down control in the Serengeti ecosystem is continually occurring. Nkwabi et al. [100] observed that insectivorous birds appear to limit insect numbers in grasslands and predicted that any removal of birds that feed on insects along agricultural land could result in an increased insect population which are pests for crops, and could become a problem for agricultural food production.

Another evidence reported by Byrom et al. [101] who studied the importance of black-winged kites formerly was black-shouldered kites (*Elanus caeruleus*) and the study was related to rodent pest control. They suggested that raptors in natural Savanna control the numbers of rodents that are pests for agricultural grain stores. In agricultural areas, these raptors are almost absent and rodent populations are six times higher than those in natural Savannas [101]. Thus, the absence of top trophic level birds could result in a pest problem eruption of rodents. The increase in rodents will result in food chain collapse due to pest control problems which will face farmers in communal lands that will lead to decreased food storage.

These results suggested that any decrease in insectivore and herbivore bird species diversity in Serengeti National Park will result not from a decline in the number of species but from the very large numbers of some individual populations, particularly those of granivorous such as queleas, bishops, widowbirds, and whydahs. These numbers have reported to be resulted in lower diversity estimates, although the overall richness was higher in Savanna.

The lower species diversity and richness of vertebrate feeders in agricultural land were probably due to increased human activities including deforestation, overgrazing, and increase in pesticide use, poisoning of predators, overhunting, and direct harassment. This finding agrees with the hypothesis that top-feeding guild bird species are less able to tolerate change in habitat compared to species at lower feeding guilds. Generally, disturbance becomes magnified up the feeding guild so that birds that feed on vertebrates experience disturbance more than do herbivores [102, 103]. It has been shown that agricultural activities are among the factors that influence raptor species richness and abundance, raptor numbers within 30 km of protected area boundaries were 40% lower than those in the interior protected area [104]. Agriculture also has an effect on populations inside protected areas close to boundaries [103, 105, 106].

The decrease in species richness of herbivorous and insectivorous birds in agricultural land was probably due to a reduced number and diversity of native tree, shrub and herb species, and the diversity and abundance of invertebrates which are a source of food for most bird species, as has been shown around Serengeti National Park [1]. In addition, a few granivorous such as queleas and widowbirds have increased in abundance by taking advantage of the grain crops. Hulme and Cresswell [107] observed that agricultural habitat with monocultures of cereal crops like maize, sorghum, and millet was more attractive to some bird species than the woodland and shrubs of native Savanna around Jos in Nigeria. In contrast, Dhindsa and Saini [108] reported an increase in the richness of some granivorous bird species in the agricultural habitats in India.

Finally, the foraging efficiency of insectivorous and herbivorous birds can be influenced by predation risk due to changes in vegetation structure and abundance. It has been suggested that increased predation rates may result in declines of some bird species in agricultural land [109, 110].

6. Conclusion and recommendation

The main objective of the study was to examine the effects of agriculture on two aspects of the Savanna bird community. First, how did agriculture affect species that differed in range within Africa? The study aimed at finding out whether birds that are endemic, regional, and Africa-wide ranges differed in their sensitivity to habitat change due to agriculture. Second, how did agriculture affect species at different trophic levels? The study tested the prediction that birds at top feeding levels (vertebrate feeders) were less able to tolerate change in habitat compared to those species at lower feeding guilds (insectivores and herbivores). The results showed a greater proportional loss of endemic species compared to wider ranging species when the habitat was changed to agriculture. In general, local endemic bird species in the Serengeti ecosystem showed less tolerance to habitat change in agricultural land.

Secondly, top feeding guilds lost proportionately more species than lower feeding guilds when habitat changed to agriculture. Thus, vertebrate feeders were less able to tolerate changes in habitat structure in agricultural land compared to the lower trophic levels of insectivores and herbivores. This result is important because there is a known top-down control by predators on lower feeding guilds in the food chain. Vertebrate feeders appear to limit rodent numbers [101] and insectivorous birds control insects [111] in Savanna compared to agriculture. Therefore, the disappearance of vertebrate feeders and insectivores could result in higher levels of rodents and some insect pests in agricultural land. The effect of food chain collapse becomes a pest control problem facing farmers.

Findings from this study imply that if climate change causes the system to move up the rainfall gradient, birds will possibly also move into agricultural areas and so lose the fragile species. These movements of birds in East Africa toward wetter areas (because of environmental change) have now been documented by Beale et al. [112]; thus, these losses of birds have taken place since then. If such losses of birds are typical of agricultural landscapes in East Africa, then many of the African savannah birds will become confined to protected areas in the next few decades [112] and some of these species could be threatened with extinction because they are only found in these protected areas.

Acknowledgements

This work was funded by a Canadian Natural Sciences and Engineering Research Council grant to Prof. A.R.E. Sinclair and Dr. Simon A.R. Mduma. We thank the Serengeti National Park, Department of Zoology and Wildlife Conservation of University of Dar es Salaam, Serengeti Wildlife Research Centre, Tanzania Wildlife Research Institute, Tanzania National Parks and Tanzania Commission for Science and Technology for their help and permission to conduct this research. We are grateful to Stephen Makacha, John Mchetto, Joseph Masoy, and Dr Bukombe John for their assistance with fieldwork.

Conflict of interest

The authors declare no conflict of interest.

Author details

Ally K. Nkwabi* and Pius Y. Kavana
Western Wildlife Research Centre, Tanzania Wildlife Research Institute, Kigoma, Tanzania

*Address all correspondence to: josenkwabi@googlemail.com

References

[1] Sinclair ARE, Mduma SAR, Arcese P. Protected areas as biodiversity benchmarks for human impact: agriculture and the Serengeti avifauna. Proceedings of the Royal Society of London. Series B: Biological Sciences. 2002;**269**(1508):2401-24

[2] Wunderlich SM, Martinez NM. Conserving natural resources through food loss reduction: production and consumption stages of the food supply chain. International Soil and Water Conservation Research. 2018;**6**(4):331-339

[3] Singh RL, Singh PK. Global environmental problems. Principles and Applications of Environmental Biotechnology for a Sustainable Future. 2017:13-41

[4] Wang L, Vo XV, Shahbaz M, Ak A. Globalization and carbon emissions: is there any role of agriculture value-added, financial development, and natural resource rent in the aftermath of COP21? Journal of Environmental Management. 2020;**268**:110712

[5] Moorcroft PR, Levin SA. Biodiversity patterns in managed and natural landscapes. In: Levin SA, Carpenter SR, Godfray HCJ, Kinzig AP, Loreau M, Losos JB, Walker B, Wilcove DS, editors. The Princeton Guide to Ecology. Princeton, New Jersey.: Princeton University Press; 2009. pp. 445-457

[6] Reside AE, Beher J, Cosgrove AJ, Evans MC, Seabrook L, Silcock JL, et al. Ecological consequences of land clearing and policy reform in Queensland. Pacific Conservation Biology. 2017;**23**(3):219-230

[7] Saridnirun G, Sukumal N, Grainger MJ, Savini T. Living with human encroachment: status and distribution of Green Peafowl in northern stronghold of Thailand. Global Ecology and Conservation. 2021;**28**:e01674

[8] Bukombe JK, Nkwabi AK, Mangewa LJ, Sweke EA, Kavana PY, Liseki SD, et al. Alien invasive species in Tanzania. In: Pullaiah T, Ielmini M, editors. Invasive Alien Species: Observations and Issues from Around the World. Wiley-Blackwell: USA; 2021. pp. 263-290

[9] Estes AB, Kuemmerle T, Kushnir H, Radeloff VC, Shugart HH. Land-cover change and human population trends in the greater Serengeti ecosystem from 1984-2003. Biological Conservation. 2012;**147**(1):255-263

[10] Estes AB, Kuemmerle T, Kushnir H, Radeloff VC, Shugart HH. Agricultural expansion and human population trends in the greater Serengeti ecosystem from 1984-2003. In: Sinclair ARE, Metzger KL, Fryxell JM, Mduma SAR, editors. Serengeti IV: Sustaining Biodiversity in a Coupled Human-Natural System. Chicago: University of Chicago Press; 2015. pp. 513-531

[11] Kideghesho JR, Nyahongo JW, Hassan SN, Tarimo TC, Mbije NE. Factors and ecological impacts of wildlife habitat destruction in the Serengeti ecosystem in northern Tanzania. African Journal of Environmental Assessment and Management. 2006;**11**:17-32

[12] Sinclair A, Hopcraft JGC, Olff H, Mduma SA, Galvin KA, Sharam GJ. Historical and future changes to the Serengeti ecosystem. In: Sinclair ARE, Packer C, Mduma SA, Fryxell JM, editors. Serengeti III: Human Impacts on

Ecosystem Dynamics. Chicago: Chicago Press; 2008. pp. 7-46

[13] Walther BA, Van Niekerk A, Rahbek C. Long-term population declines of Palearctic passerine migrant birds: a signal from the Sahel. In: British Ornithologists' Union's (BOU) Annual Conference. 2011. http://www.bou.org.uk/bouproc-net/migratory-birds/walther-etal.pdf. [Accessed: 12 December 2013]

[14] Sinclair AR, Nkwabi A, Mduma SA, Magige F. Responses of the Serengeti avifauna to long-term change in the environment. Ostrich. 2014;**85**(1):1-11

[15] Mkumbukwa AR. The evolution of wildlife conservation policies in Tanzania during the colonial and post-independence periods. Development Southern Africa. 2008;**25**(5):589-600

[16] Ogutu JO, Owen-Smith N, Piepho HP, Said MY. Continuing wildlife population declines and range contraction in the Mara region of Kenya during 1977-2009. Journal of Zoology. 2011;**285**(2):99-109

[17] Nkwabi AK, Bukombe J, Maliti H, Liseki S, Lesio N, Kija H. An overview of biodiversity of tanzania and conservation efforts. In: Pullaiah T, editor. Global Biodiversity Volume 3, Selected Countries in Africa. Waretown, USA.: Apple Academic Press; 2018. pp. 297-342

[18] Kauzeni AS, Kiwasila HL. Serengeti Regional Conservation Strategy: A Socio-Economic Study. Tanzania: Institute of Resource Assessment, University of Dar es Salaam; 1994

[19] Norton-Griffiths M, Herlocker D, Pennycuick L. The patterns of rainfall in the Serengeti ecosystem, Tanzania. East African Wildlife Journal. 1975;**13**(3-4):347-374

[20] Sinclair ARE, Mduma SAR, Hopcraft JGC, Fryxell JM, Hilborn R, Thirgood S. Long-term ecosystem dynamics in the Serengeti: lessons for conservation. Conservation Biology. 2007;**21**(3):580-590

[21] Sinclair ARE, Mduma SAR, Hopcraft JGC, Galvin K, Sharam GJ. Historical and future changes to the Serengeti ecosystem. In: Sinclair ARE, Packer C, Mduma SAR, Fryxell JM, editors. Serengeti III: Human Impacts on Ecosystem Dynamics. Chicago: University of Chicago Press; 2008. pp. 7-46

[22] Morell V. Counting creatures of the Serengeti, great and small. Science. 1997;**278**(5346):2058-2060

[23] Estes A. Avian Diversity in Serengeti National Park: How do Community Differ Between Interior Savanna and Grassland and Exterior Agricultural Landscape [MSc Thesis]. Madison: University of Wisconsin. 2000

[24] Sekercioglu CH. Increasing awareness of avian ecological function. Trends in Ecology and Evolution. 2006;**21**(8):464-471

[25] Sinclair ARE, Metzger KL, Fryxell JM, Packer C, Byrom A, Craft ME, et al. Asynchronous food-web pathways could buffer the response of Serengeti predators to El Niño Southern Oscillation. Ecology. 2013;**94**(5):1123-1130

[26] Metzger KL, Sinclair ARE, Macfarlane A, Coughenour MB, Ding J. Scales of Change in the Greater Serengeti Ecosystem. In: Sinclair ARE, Metzger KL, Mduma SAR, editors. Serengeti IV: Sustaining Biodiversity in a Coupled Human-Natural System. Chicago: University of Chicago Press; 2015. pp. 33-71

[27] Ikin K, Beaty RM, Lindenmayer DB, Knight E, Fischer J, Manning AD. Pocket parks in a compact city: how do birds respond to increasing residential density? Landscape Ecology. 2013;**28**:45-56

[28] Newbold T, Scharlemann JP, Butchart SH, Şekercioğlu ÇH, Alkemade R, Booth H, et al. Ecological traits affect the response of tropical forest bird species to land-use intensity. Proceedings of the Royal Society B: Biological Sciences. 2013;**280**(1750):2012-2131

[29] Sykes MT, Prentice IC, Cramer W. A bioclimatic model for the potential distributions of north European tree species under present and future climates. Journal of Biogeography. 1996;**23**(2):203-233

[30] Pearson RG, Dawson TP. Predicting the impacts of climate change on the distribution of species: are bioclimate envelope models useful? Global Ecology and Biogeography. 2003;**12**(5):361-371

[31] Both C, Visser ME. The effect of climate change on the correlation between avian life-history traits. Global Change Biology. 2005;**11**(10):1606-1613

[32] Williams P, Hannah L, Andelman S, Midgley G, AraúJo M, Hughes G, et al. Planning for climate change: identifying minimum-dispersal corridors for the Cape Proteaceae. Conservation Biology. 2005;**19**(4):1063-1074

[33] Burger C, Belskii E, Eeva T, Laaksonen T, Mägi M, Mänd R, et al. Climate change, breeding date and nestling diet: how temperature differentially affects seasonal changes in pied flycatcher diet depending on habitat variation. Journal of Animal Ecology. 2012;**81**(4):926-936

[34] Grainger A. Constraints on modelling the deforestation and degradation of tropical open woodlands. Global Ecology and Biogeography. 1999;**8**(3-4):179-190

[35] Homewood K, Brockington D. Biodiversity, conservation and development in Mkomazi Game Reserve, Tanzania. Global Ecology and Biogeography. 1999;**8**(3-4):301-313

[36] Thomas CD, Gillingham PK. The performance of protected areas for biodiversity under climate change. Biological Journal of the Linnean Society. 2015;**115**(3):718-730

[37] Thomas CD, Cameron A, Green RE, Bakkenes M, Beaumont LJ, Collingham YC, et al. Extinction risk from climate change. Nature. 2004;**427**(6970):145-148

[38] Sekercioglu CH, Primack RB, Wormworth J. The effects of climate change on tropical birds. Biological Conservation. 2012;**148**(1):1-18

[39] Cowling R, Pressey R, Lombard A, Desmet P, Ellis A. From representation to persistence: requirements for a sustainable system of conservation areas in the species-rich mediterranean-climate desert of southern Africa. Diversity and Distributions. 1999;**5**(1-2):51-71

[40] Rutherford M, Powrie L, Schulze R. Climate change in conservation areas of South Africa and its potential impact on floristic composition: a first assessment. Diversity and Distributions. 1999;**5**(6):253-262

[41] Araújo MB, Cabeza M, Thuiller W, Hannah L, Williams PH. Would climate change drive species out of reserves? An assessment of existing reserve-selection methods. Global Change Biology. 2004;**10**(9):1618-1626

[42] Gill JA. Approaches to measuring the effects of human disturbance on birds. Ibis. 2007;**149**(Suppl. 1):9-14

[43] Hole DG, Willis SG, Pain DJ, Fishpool LD, Butchart SH, Collingham YC, et al. Projected impacts of climate change on a continent-wide protected area network. Ecology Letters. 2009;**12**(5):420-431

[44] Carey C. The impacts of climate change on the annual cycles of birds. Philosophical Transactions of the Royal Society B: Biological Sciences. 2009;**364**(1534):3321-3330

[45] Sillett TS, Holmes RT, Sherry TW. Impacts of a global climate cycle on population dynamics of a migratory songbird. Science. 2000;**288**(5473):2040-2042

[46] Myers N, Knoll AH. The biotic crisis and the future of evolution. Proceedings of the National Academy of Sciences. 2001;**98**(10):5389-5392

[47] Walker BH. Biodiversity and ecological redundancy. Conservation Biology. 1992;**6**(1):18-23

[48] Walker BH. Conserving biological diversity through ecosystem resilience. Conservation Biology. 1995;**9**(4):747-752

[49] Rapport D, Costanza R, McMichael A. Assessing ecosystem health. Trends in Ecology & Evolution. 1998;**13**(10):397-402

[50] Chapin FS, Zavaleta ES, Eviner VT, Naylor RL, Vitousek PM, Reynolds HL, et al. Consequences of changing biodiversity. Nature. 2000;**405**(6783):234-242

[51] May RM. Thresholds and breakpoints in ecosystems with a multiplicity of stable states. Nature. 1977;**269**(5628):471-477

[52] Holling CS. Resilience and stability of ecological systems. Annual Review of Ecology and Systematics. 1973;**4**:1-23

[53] Cannell M, Thornley J, Mobbs D, Friend A. UK conifer forests may be growing faster in response to increased N deposition, atmospheric CO_2 and temperature. Forestry. 1998;**71**(4):277-296

[54] Pavlacky DC, Possingham HP, Lowe AJ, Prentis PJ, Green DJ, Goldizen AW. Anthropogenic landscape change promotes asymmetric dispersal and limits regional patch occupancy in a spatially structured bird population. Journal of Animal Ecology. 2012;**81**(5):940-952

[55] Parmesan C. Climate and species' range. Nature. 1996;**382**(6594):765-766

[56] Jezkova T. The Ecological Niche and Ecological Niche Modeling. Las Vegas: School of Life Sciences, University of Nevada; 2012

[57] Mathews S, Bonser SP. Life histories, ecological tolerance limits, and the evolution of geographic range size in Eucalyptus (Myrtaceae). Australian Journal of Botany. 2005;**53**(6):501-508

[58] Gilchrist GW. Specialists and generalists in changing environments. I. Fitness landscapes of thermal sensitivity. American Naturalist. 1995;**146**(2):252-270

[59] Gaston KJ. Geographic range limits of species. Proceedings of the Royal Society B: Biological Sciences. 2009;**276**(1661):1391-1393

[60] Nkwabi AK, Bukombe JK, Kija HK, Liseki SD, Ndimuligo SA,

Kavana PY. Avifauna in relation to habitat disturbance in Wildlife Management Areas of the Ruvuma miombo ecosystem, Southern Tanzania. In: Mikkola H, editor. Birds - Challenges and Opportunities for Business, Conservation and Research. London, UK, London: IntechOpen; 2021. pp. 1-20

[61] Nkwabi AK, John B, Kija H, Otsyina RM, Monjare JF, Kajuni AR. Abundance and distribution of small mammals relative to human activities in the wildlife management areas of Ruvuma landscape, southern Tanzania. International Journal of Fauna and Biological Studies. 2018;**5**(2):156-162

[62] Thonfeld F, Steinbach S, Muro J, Hentze K, Games I, Naeschen K, et al. The impact of anthropogenic land use change on the protected areas of the Kilombero catchment, Tanzania. ISPRS Journal of Photogrammetry and Remote Sensing. 2020;**168**:41-55

[63] Kideghesho JR, Rija A, Mwamende K, Selemani I. Emerging issues and challenges in conservation of biodiversity in the rangelands of Tanzania. Nature Conservation. 2013;**6**:1-29. DOI: 10.3897/natureconservation.6.5407

[64] Mwalyosi RB. Land-use Changes and Resource Degradation in South-West Masailand, Tanzania. Environmental Conservation. 1992;**19**(02):145-152

[65] Rybicki J, Hanski I. Species–area relationships and extinctions caused by habitat loss and fragmentation. Ecology Letters. 2013;**16**(Suppl.1):27-38

[66] Kideghesho JR, Kaltenborn BP, Røskaft E, Tarimo TM. Serengeti shall not die: can the ambition be sustained? The International Journal of Biodiversity Science and Management. 2005;**1**(3):150-166

[67] Sinclair ARE, Norton-Griffiths M, editors. Serengeti: Dynamics of an Ecosystem. Chicago: University of Chicago Press; 1979

[68] Sinclair I, Ryan P. Birds of Africa: South of the Sahara. Cape Town: Random House, Struik; 2010

[69] Fry CH, Keith S, Urban EK, editors. Birds of Africa Volume VI. Picathartes to Oxpeckers. San Diego, California: Academic Press; 2000

[70] Fry CH, Keith S. The birds of Africa Volume VII. Sparrows to buntings. London: Christopher Helm; 2004

[71] Simberloff D, Dayan T. The guild concept and the structure of ecological communities. Annual Review of Ecology and Systematics. 1991;**22**:115-143

[72] Hammer Ø, Harper DAT, Ryan PD. PAST: paleontological statistics software package for education and data analysis. Palaeontologia Electronica. 2001;**4**(1):1-9

[73] Krebs CJ, Kenney AJ. Programs for Ecological Methodology. Menlo Park: Bennjamin/Cummings; 2009

[74] Krebs CJ. Ecological Methodology. New York, NY: Harper and Row; 2001

[75] Seaby RMH, Henderson PA. Species Diversity and Richness Version 4.1.2. 2007, Pisces Conservation Ltd: Lymington

[76] Solow AR. A simple test for change in community structure. Journal of Animal Ecology. 1993;**62**(1):191-193

[77] Terborgh J, Lopez L, Nunez P, Rao M, Shahabuddin G, Orihuela G, et al. Ecological meltdown in predator free forest fragments. Science. 2001;**294**(5548):1923-1926

[78] Bonier F, Martin PR, Wingfield JC. Urban birds have broader environmental tolerance. Biology Letters. 2007;**3**(6):670-673

[79] Waltert M, Mardiastuti A, Mühlenberg M. Effects of land use on bird species richness in Sulawesi, Indonesia. Conservation Biology. 2004;**18**(5):1339-1346

[80] Bolwig S, Pomeroy D, Tushabe H, Mushabe D. Crops, trees, and birds: biodiversity change under agricultural intensification in Uganda's farmed landscapes. Geografisk Tidsskrift, Danish Journal of Geography. 2006;**106**(2):115-130

[81] Neuschulz EL, Botzat A, Farwig N. Effects of forest modification on bird community composition and seed removal in a heterogeneous landscape in South Africa. Oikos. 2011;**120**(9):1371-1379

[82] Van Dorp D, Opdam P. Effects of patch size, isolation and regional abundance on forest bird communities. Landscape Ecology. 1987;**1**(1):59-73

[83] Lepczyk CA, Flather CH, Radeloff VC, Pidgeon AM, Hammer RB, Liu J. Human impacts on regional avian diversity and abundance. Conservation Biology. 2008;**22**(2):405-416

[84] Soini E. Bird diversity and land use on the slopes of Mt Kilimanjaro and the adjacent plains, Tanzania. African Zoology. 2006;**41**(2):193-209

[85] Newmark WD. Isolation of African protected areas. Frontiers in Ecology and the Environment. 2008;**6**(6):321-328

[86] Devictor V, Godet L, Julliard R, Couvet D, Jiguet F. Can common species benefit from protected areas? Biological Conservation. 2007;**139**(1):29-36

[87] Sheehan DK, Sanderson FJ. Seeing the bigger picture: how anthropogenic landscape modification in Africa affects declining migratory birds and the need for trans-continental research and conservation. Ibis. 2012;**154**(4):659-662

[88] Moreau RE. Mt Kilimanjaro and Mt Kenya: some comparisons with special reference to mammals and birds with a note on Mt Meru (with bibliography of Kilimanjaro by C. Gillman). Tanganyika Notes and Records. 1944;**14**:28-59

[89] Misana SB, Majule AE, Lyaruu HV, Change LU. Linkages between changes in land use, biodiversity and land degradation on the slopes of Mount Kilimanjaro, Tanzania. LUCID Project, International Livestock Research Institute. 2003

[90] Atkinson PW, Buckingham D, Morris AJ. What factors determine where invertebrate-feeding birds forage in dry agricultural grasslands? Ibis. 2004;**146**(Suppl. 2):99-107

[91] Long P, Székely T, Kershaw M, O'Connell M. Ecological factors and human threats both drive wildfowl population declines. Animal Conservation. 2007;**10**(2):183-191

[92] Woodroffe R, Ginsberg JR. Edge effects and the extinction of populations inside protected areas. Science. 1998;**280**:2126-2128

[93] Barnes AD, Allen K, Kreft H, Corre MD, Jochum M, Veldkamp E, et al. Direct and cascading impacts of tropical land-use change on multi-trophic biodiversity. Nature Ecology & Evolution. 2017;**1**(10):1511-1519

[94] Rurangwa ML, Aguirre-Gutiérrez J, Matthews TJ, Niyigaba P, Wayman JP, Tobias JA, et al. Effects of land-use change on avian taxonomic, functional

and phylogenetic diversity in a tropical montane rainforest. Diversity and Distributions. 2021;**27**(9):1732-1746

[95] Müller S, Gossner MM, Penone C, Jung K, Renner SC, Farina A, et al. Land-use intensity and landscape structure drive the acoustic composition of grasslands. Agriculture, Ecosystems & Environment. 2022;**328**:107845

[96] Sinclair ARE. Population regulation in animals, ecological concepts. Symposium of the British Ecological Society. 1989;**29**:197-241

[97] Sinclair ARE, Metzger K, Brashares JS, Nkwabi A, Sharam G, Fryxell JM, et al. Trophic cascades in African savanna: Serengeti as a case study. In: Terborgh J, Estes J, editors. Trophic Cascades: Predators, Prey, and the Changing Dynamics of Nature. Washington, DC.: Island Press; 2010. pp. 255-274

[98] Terborgh J, Estes JA, editors. Trophic Cascades: Predators, Prey, and the Changing Dynamics of Nature. 2010

[99] Estes JA, Terborgh J, Brashares JS, Power ME, Berger J, Bond WJ, et al. Trophic downgrading of planet earth. Science. 2011;**333**(6040):301-306

[100] Nkwabi AK, Sinclair ARE, Metzger KL, Mduma SAR. The effect of natural disturbances on the avian community of the Serengeti woodlands. In: Sinclair ARE, Metzger K, Mduma SAR, Fryxell J, editors. Serengeti IV: Sustaining Biodiversity in a Coupled Human-Natural System. Chicago: University of Chicago; 2015. pp. 395-418

[101] Byrom AE, Ruscoe W, Nkwabi AJK, Metzger K, Forrester GJ, Durant S, et al. Small mammal diversity and population dynamics in the greater Serengeti ecosystem. In: Sinclair ARE, Metzger K, Mduma SAR, Fryxell J, editors. Serengeti IV: Sustaining Biodiversity in a Coupled Human-Natural System. Chicago: University of Chicago; 2015. pp. 323-357

[102] Tuule E, Tuule A, Lõhmus A. Fifty-year dynamics in a temperate raptor assemblage. Estonian Journal of Ecology. 2011;**60**:132-142

[103] Anadón JD, Sánchez-Zapata JA, Carrete M, Donázar JA, Hiraldo F. Large-scale human effects on an arid African raptor community. Animal Conservation. 2010;**13**(5):495-504

[104] Herremans M, Herremans-Tonnoeyr D. Land use and the conservation status of raptors in Botswana. Biological Conservation. 2000;**94**(1):31-41

[105] Thiollay JM. The decline of raptors in West Africa: long-term assessment and the role of protected areas. Ibis. 2006;**148**(2):240-254

[106] Ogada DL, Keesing F. Decline of raptors over a three-year period in Laikipia, central Kenya. Journal of Raptor Research. 2010;**44**(2):129-135

[107] Hulme MF, Cresswell W. Density and behaviour of Whinchats Saxicola rubetra on African farmland suggest that winter habitat conditions do not limit European breeding populations. Ibis. 2012;**154**(4):680-692

[108] Dhindsa MS, Saini HK. Agricultural ornithology: an Indian perspective. Journal of Biosciences. 1994;**19**(4):391-402

[109] Whittingham MJ, Evans KL. The effects of habitat structure on predation risk of birds in agricultural landscapes. Ibis. 2004;**146**(Suppl. 2):210-220

[110] Marsden SJ, Symes CT, Mack AL. The response of a New Guinean avifauna

to conversion of forest to small-scale agriculture. Ibis. 2006;**148**(4):629-640

[111] Nkwabi A, Sinclair AE, Metzger K, Mduma S. Disturbance, species loss and compensation: wildfire and grazing effects on the avian community and its food supply in the Serengeti Ecosystem, Tanzania. Austral Ecology. 2011;**36**(4):403-412

[112] Beale CM, Baker NE, Brewer MJ, Lennon JJ. Protected area networks and savannah bird biodiversity in the face of climate change and land degradation. Ecology Letters. 2013;**16**(8):1061-1068

Chapter 7

The Structure and Functions of the Contour Feathers of Water Birds Revisited

Arie M. Rijke

Abstract

The structural details of the contour feathers of water birds are known to serve a variety of functions ranging from intraspecific signaling to such physical qualities as thermal insulation, water repellency, resistance to impact and viscous drag reduction. All of them closely match the requirements of habitat and feeding habits. It comes as no surprise, therefore, that contour feathers are composed of an array of elements that confer these qualities to the optimal benefit of their avian bearer. In this chapter, we review the physical bases for these functions expressed in both structural and evolutionary terms. Some functions excel at the expense of others and many species have evolved an optimal balance between feather micro-structure and behavioral patterns that suit their specific environment. Several relationships between feather function and the structural properties of water bird feathers can be identified as specific evolutionary adaptations.

Keywords: water birds, feather structure, behavioral patterns, feather adaptations, contour feather function

1. Introduction

Water birds belong to a large group of families that have open water in their habitat. They make their home in many diverse environments including oceans, shores, estuaries, rivers, lakes and wetlands. Many of them feed in salt water, taking their prey from the surface or catching it under water by swimming, plunging and deep diving. Some exploit the skies above pursuing their prey in an unobstructed three-dimensional space without ever alighting. Others inhabit only fresh water habitats and forage in a wide variety of ways such as skimming the surface or stalking under water Among them are families that have colonized the remotest parts of our earth and have adapted to the most extremes of climatic conditions. Indeed, water birds can be found foraging and often breeding at all latitudes. They can truly be said to have conquered the entire aquatic world.

Such wide-spread occurrence has exposed water birds to a multitude of evolutionary forces that have shaped their anatomy and behavioral patterns to optimally suit their specific environment. In this chapter, we will show how the feathers of water birds, in particular the contour feathers, vary among families and exhibit a range of properties that function, among other things, to regulate body temperature, repel water, prevent water from penetrating to the skin and resist the impact forces of diving, plunging and alighting. Some of these functions excel in extreme environmental conditions or in relation to specific feeding techniques, frequently at the expense of other functions. Others represent a balance between two or more opposing functions. In consequence, many of these functions are expressed in an array of feather characters that confer these qualities to the optimal benefit of the avian bearer.

To study the relationship between structure and function in some detail, a closer look at the composition of feathers is in order. The morphology of feathers has been well described in the ornithological literature [1, 2] and is reproduced here only to the extent necessary for the purpose of this chapter. All feathers, whether flight, tail or contour feathers, consist of essentially the same elements, only their relative prominence is different. At the base of the spine (or rachis) occur the downy or plumulaceous feathers, only a tuft in flight and tail feathers, but extensively present alongside the proximal two-third of the rachis of contour feathers. These are thought to primarily function as a means to regulate body temperature by entrapping air [3, 4]. More distally, a highly structured pattern is present with rami extending from the rachis in the plane of the feather, each sprouting barbules of which the distal ones have hooks that catch upon the curled proximal barbules of the barb next more distal. This continuous-looking, hook-and-flange arrangement provides this pennaceous part of the feather with the rigidity so critical for its mechanical and other properties.

In flight and tail feathers, the pennaceous part is by far the most dominant part – reflecting their importance in flight - but in contour feathers it occupies only the distal one-third of the length of the feather. The proximal two-third is made up of downy elements that work, among other things, as structural reinforcements limiting the bending of the downy barbules.

Efforts to correlate structure and function require an exact knowledge of all respective feather elements and a full understanding of their specific functions. However, much as we know about feather anatomy, not all details can be explained in terms of functional performance. For instance, it has been well established that suites of microscopic characters seen in downy barbules can be used to identify certain groups of birds [5] and, more recently, to correlate habitat and behavioral patterns with variations observed in these microscopic characters [6]. Because downy feathers are thought to primarily function as a means to regulate body temperature by entrapping air, the presence in the down of any species- or group-specific character suggests that environmental forces other than thermoregulation can and do exact evolutionary changes in down. We do not know what these forces are.

We also do not know all functions of feathers or their components. However, we will address here the functions that we do know and try to correlate these with the constituent feather elements, quantitatively or semi-quantitatively where possible.

2. Signaling

Birds signal to each other by behavior and the coloration of their feathers. Color can serve, among other things, to hide from predators, to attract mates or fight off rivals and is brought about by pigment cells within the feather. Melanin is an important pigment produced by melanocytes arrayed within the barbs and barbules. Another source of color is iridescence, a nano-structurally colored sheen on the feathers of several lineages of birds [7]. As far as we know, neither pigment nor iridescence interferes directly with the functionality of other feather elements.

3. Thermoregulation

As mentioned earlier, downy feathers are thought to function by entrapping air. The downy feather characters, including the nodes observed in many water bird and other families, are believed to work as structural reinforcements that limit the bending of the downy barbules. They may also function to catch other barbules and keep them from becoming entangled, thereby allowing the entrapment of more air and serve as a better thermal insulator [8, 9].

If the presence of nodes and other downy characteristics contribute to the conservation of heat by air convection, one may expect additional adaptations in the down of contour feathers for the conservation of body heat radiation. The nodes with their high content of dark granules, most likely consisting of melanin, may trap heat through infrared absorption and function by storing heat.

The surface of a bird's skin loses energy in the form of heat through two mechanisms: one is convection heat, warmed air that is trapped by down and overlying pennaceous feathers as stagnant air, the other is heat by radiation, emitted from the skin of all warm-blooded animals as detectable by infrared cameras. Part of the radiation is absorbed by the feather keratin and, in turn, converted into convection heat and, partially re-emitted from the keratin or lost to the surrounding environment. By investigating the infrared absorption spectra of contour feathers it is possible to identify the contribution to radiation heat conservation by their respective downy and pennaceous parts. This has been done for the King Penguin (*Aptenodytes patagonicus*), the Wood Stork (*Mycteria americanus*) and the American Crow (*Corvus brachyrhynchos*) as representatives of divers, waders and land birds, respectively [10]. The results show essentially identical spectra for the three species over the range of 500 to 4500 cm^{-1} wavenumbers suggesting there are no adaptative effects of foraging niche and habitat on the radiation heat conservation mechanisms of these and possibly all species.

There is, however, a prominent difference between the spectra of downy feathers and the pennaceous parts at 1700 cm^{-1}. Here, down shows a sharp transmittance peak that is strikingly absent in the pennaceous. The occurrence of this peak can be best explained by the Christiansen effect [11, 12] observed in binary systems, such a liquid and a powder, where the transmittance peak occurs at a wavelength for which both components have the same refractive index. In the case of down and air, this occurs at 1700 cm^{-1}. At all other wavelengths, scattering occurs due to the difference in the refractive index between down and air.

The values of the refractive index of materials, and the way these depend on wavelength, are determined by several factors. Apart from chemical composition, crystallinity with its various axial orientations, contributes significantly to these values. In long-chain molecules, such as keratin, the extent of crystallinity and the preferential orientation of crystalline regions provide a level of ordered structure that contributes as well. Non-crystalline alignment of polypeptide chains due to interchain bonding such as electrostatic bonds and hydrogen bonds, can also be expected to contribute to the value of the refractive index [10].

As the infrared spectra have shown, the chemical composition of all parts of the contour feathers is essentially the same and is dominated by the bulk of the beta-keratin component. However, the refractive indices for the various wavelengths for down and pennaceous are not the same due to their differences in crystallinity and structural orientation. As a result, the radiation peak transmitted by down at 1700 cm^{-1} is not observed for the pennaceous part of the feather. Its absence is most likely explained by the limited range of wave numbers employed to coincide with the range of body heat radiation. Had the range of wave numbers been extended beyond the infrared, it is likely that a transmittance peak for the pennaceous could have been located, provided, of course, that a wavelength for which pennaceous and air have the same refractive index actually exist.

Similar considerations are expected to apply to radiation heat that originates from the environment. Here, infrared from solar radiation with wave numbers comparable to those of body heat radiation will be scattered or partly absorbed by the pennaceous feather parts and not directly reach the skin of the bird. For wave numbers extending beyond the infrared range, it is, again, possible that a transmittance peak could be located for the pennaceous portion of the feather, but the down is not likely to be permeable at that particular wave number. The difference in feather ultrastructure between the down and the pennaceous parts, with its concomitant difference in infrared transmittance, is the reason why radiation heat, whether from body heat or an environmental source, does not directly permeate the feather coat.

The Christiansen effect with its transmittance peak at 1700 cm^{-1}, as observed in the infrared spectra of down but absent in the pennaceous parts, is likely to have important consequences for the conservation of radiation heat by the bird. Heat radiated from the body will be scattered and re-scattered in all directions by the downy barbules, reflecting radiation back to the skin where it is absorbed. However, radiation at 1700 cm^{-1} will be fully transmitted by the down but not by the pennaceous, which will scatter it again, thereby conserving heat within the feather coat.

The nodes in the down, and in particular those that contain melanin, may aid in the process of radiation heat conservation by virtue of their heat capacitance but efforts to document infrared spectra on isolated nodes were unsuccessful [10]. Therefore, it has not been possible to determine whether nodes play any direct role in the conservation of body heat radiation.

The containment of body heat radiation is likely to be of secondary importance in the overall conservation of body warmth in birds. No data on total heat conductivity have been experimentally determined for the radiation part in feathers, but estimates for diabase wool used in the construction sector are near 25 percent [13]. However, because most of the radiation in birds is scattered by down, converted to convection heat and partly re-emitted at a different, higher wavelength, it is evident that the transmittance leak at 1700 cm^{-1} constitutes only a small loss of thermal energy by the down. The actual loss is probably minimal as this transmittance is scattered back to the skin by the pennaceous parts of the contour feather.

If the Christiansen effect is insignificant in the loss of radiation heat, then one can speculate whether the transmittance peak at 1700 cm^{-1} and its deflection by the overlying pennaceous parts constitute an incidental physical phenomenon or represent an adaptation to the overall thermal balance. The available data do not provide an obvious answer. However, the absence of any feather components with specific infrared properties that stop the transmittance leak seems to suggest an incidental phenomenon rather than an evolutionary adaptation. No specific infrared absorption bands in either the keratin or the pigmented nodes have been observed that could account for such a stoppage.

The amounts of radiation heat that birds generate or lose are unknown and do not appear to be amenable to direct experimental determination, at least not by current technologies. Infrared spectra of feather components have demonstrated the mechanisms by which radiation heat is conserved and lost, but have not provided for a quantitative analysis.

The distal, pennaceous part of the contour feathers is not only exposed to environmental radiation but also to air and water. It is at this interface that the physical interaction with the external world occurs and where most adaptations to environmental factors can be found.

4. Water repellency and resistance to water penetration

One of the major functions of feathers is to prevent water from reaching the skin or weighing down the remiges and tail feathers in flight. With very few exceptions, all birds benefit from a plumage that optimally repels and resists the penetration of water. However, the manner in which this optimum is realized for each water bird family is closely associated with its behavior and interaction with its habitat and, as a result, the feather characters responsible for water repellency and resistance vary accordingly. To understand the way a water repellent/resistant structure functions, certain aspects of surface physics should be made clear.

The water repellency of feathers and other biological porous structures, such as the stomatal apparatus of leaves and the spiracles of insects, is governed by the fundamental principles of surface physics that apply to all porous surfaces whether natural or manmade. It is determined by the relative areas of solid – water and air – water interface and their respective interfacial energies regardless of the actual architecture of the repellent structure itself [14]. If the surface of the solid is coated with another material, such as paint or preening oil, it will assume the properties of the coating material. For feathers coated with uropygial gland oil, the feather – water interface is, in fact, an interface between gland oil and water.

A drop of water placed on a smooth feather surface such as the rachis, will pearl up and roll off easily. This surface is then said to be water repellent, the actual extent of which is determined by the contact angle θ, defined as the angle between the tangent to the curved water surface at the point of contact with the feather surface and the plane of the surface on which the drop is resting, measured through the water. When the drop is placed on the porous vane of the feather, it will entrap air in the hollows and interstices, forming additional air – water interfaces, which will cause considerable increase in the contact angle.

The manner in which this increase occurs has been described in previous publications [14–16] and need not be repeated here. Suffice to say that the contact angle and thus the water repellency of the distal one-third of the contour feather is quantitatively determined by $(r + d)/r$ where $2r$ represents the diameter of the barbs measured in the plane of the long axes of the barbs separated by distance $2d$. Note that the increase in contact angle is ascertained by the parameter $(r + d)/r$ only and not by the separate values of r and d. Thus, θ for values of this parameter ranging between 2.4 (penguins, *Spheniscidae*) and 10 (land birds) would vary between about 126° and 154°, roughly correct by experimental verification [15]. These values are significantly higher than those attained for the most repellent of smooth surfaces which equal about 114° [16].

These results have been derived solely from basic physico-chemical principles without reference to any specific dimensions of the porous surface. They are determined only by the areas of feather – water and air – water interface per unit of macroscopic surface areas without dictating the shape, curvature or configuration of these interfaces. Therefore, the relationship between the dimensions of a porous surface and its ensuing contact angle is a rigorous one, not an empirical one, and is of general validity. These premises have been tested experimentally and were found to be correct by Cassie and Baxter and Rijke using paraffinated stainless steel wire cages and grids [14, 17]. Many other studies including recent ones, have reported contact angle measurements on porous substrates including feathers and consistently confirmed the correctness of the above premises [16–19].

In order to measure contact angles on smooth or porous surfaces correctly, certain experimental conditions have to be met, such as: the drop has to be small enough so as not to be perturbed by gravitational forces, but large enough to cover a representative area of the porous surface. The drop should be prevented from evaporation which would turn the advancing contact angle into a receding one. Feather specimens should be covered with fresh preening oil, not rinsed with an ethanol wash [18]. When these conditions are met, the correct contact angle is usually found to be within one-degree error as observed by multiple authors [14, 17, 20–23]. These results have shown conclusively that contact angles and therefore water repellencies can be reliably calculated from and represented by the dimensions of the porous surface alone.

An expression for the pressure (P), required to force water between the barbs and barbules, can be derived from similar premises and shows P to be inversely proportional to r and $(r + d)/r$. As a result, the requirement of relatively large values for $(r + d)/r$ to provide sufficient water repellency is opposed by the need for small values for this parameter to attain good resistance to water penetration. Thus, the structural characteristics compatible with optimal water repellency are, at least in part, in conflict with the requirements of resistance to water penetration. This conflict has important implications for water birds, which must realize a balance between these two opposing functions to cope with their respective habitats and behavioral patterns as indeed they do [15].

Experimental data on water repellency and resistance to water penetration for Double-crested Cormorants (*Phalacrocorax auritus*) and Anhingas (*Anhingidae*) have shown that results can be satisfactorily interpreted in terms of barb diameter and spacing only without recourse to barbules. Their $(r + d)/r$ values for barbules are in the approximate range of 4.5 to 5.5 as found for almost all bird families regardless of their feeding habits or interaction with open water [15]. This suggests that the contribution of barbules to water resistance is real, but not based on the same mechanism as applies to barbs. Barbules provide an interlocking mechanism by preventing the barbs from separating under the increasing water pressure while increasing their own separation by their hooks sliding in the flanges, a process that can be verified under a low-powered light microscope. Similarly, water drops being repelled by the barbs and not involving the barbules can be observed with a magnifying glass.

The contact angle θ of water drops on smooth feather surfaces, such as the rachis or on a microscopic slide covered with preening oil, measures about 90° as established by various authors [14, 16, 23]. The same value was found for water drops on polyethylene foil [20] and this is no coincidence: polyethylene almost exclusively consists of methylene groups ($-CH_2-$) which are the predominant chemical component of preening oil [24, 25].

That the specific architecture of a porous surface does not enter into the calculations allows the investigator to determine the contact angle from the value of $(r + d)/r$ alone. For instance, Cassie and Baxter [14] found $(r + d)/r$ for their duck feathers to be 5.9, which corresponds to a θ of 147° in good agreement with their experimental value of 150°. These results, corroborated by other workers [26], have shown that for feathers coated with fresh preening oil, both the water repellency in terms of the contact angle θ and the balance between water repellency and resistance expressed by the value of (r + d)/r, can be correctly predicted from the micro-structure of the feather alone. Furthermore, the value of 5.9 for duck feathers, when compared with 4.8 for the White-breasted Cormorant (*Phalacrocorax carbo*) [27] and 7.1 for the European Starling (*Sturnus vulgaris*) [28], suggests that the duck, and probably all dabblers, are more water repellent than cormorants, but less so than starlings. On the other hand, cormorants show a superior resistance to water penetration, particularly when compared with starlings.

Measurements on more than 160 species of about 45 bird families [26, 27, 29–31] have shown that (r + d)/r values vary from about 2.3 for penguins to about 6.5 for gulls (*Laridae*) and up to 10 for most terrestrial birds. This range in values for this parameter suggests that each water bird family has evolved a balance between water repellency and resistance to water penetration that suits its particular habitat and behavioral pattern.

The data on barb diameter, spacing, and (r + d)/r values published in the peer-reviewed literature are far from a complete inventory of bird plumage, but on the basis of what is available, the following observations can be made and tentative conclusions reached. First, the distal one-third of breast, abdominal, and back feathers shows the patterned structure that confers the water repellency and resistance to penetration. The proximal and medial parts show no such structure. The tail feathers and remiges, on the other hand, show a pennaceous structure over essentially the entire length of the feather and have values of (r + d)/r that are generally small, which prevent these feathers from becoming waterlogged. Among water bird families, contour feathers vary more in values of (r + d)/r, which range from 2 to 10, than rectrices and remiges both of which vary little and range from 2 to 4 [14, 30, 32]. Second, within most families, the contour feathers that protect the skin from coming in contact with water have, on the whole, very similar values for (r + d)/r, exceptions seen only when a species within a family behaves differently from its relatives. A typical example is the Brown Pelican (*Pelecanus occidentalis*), which, unlike its congeners, dives for its prey from the air. Third, data sets on feather structure suggest a relationship between barb diameter 2r and (r + d)/r values. Families such as penguins and other diving water birds, have wide barb diameters and small values for (r + d)/r, whereas the opposite holds true for terrestrial families such as the starlings and nightjars (*Caprimulgidae*). Birds that come into occasional contact with open water such as herons (*Ardeidae*) and gulls have intermediate values. As a result, penguins have excellent resistance to water penetration but poor water repellency as shown by their familiar 'wet' appearance when they exit the water. The breast feathers of terrestrial birds, on the other hand, are very water repellent but promise little in the way of resistance to water penetration. Those of herons and gulls fall somewhere in between.

5. Water repellency, water resistance and behavioral patterns

The first effort to correlate the value of the parameter (r + d)/r - that is, the balance between water repellency and resistance to water penetration - with behavioral patterns was made more than 50 years ago [33]. In that paper, the well-known habit of cormorants of spreading their wings to the sun or breeze after a period in the water, a feature commonly referred to as "wing-drying," was proposed to follow from the poor water repellency of their breast feathers, as evidenced by their low value for (r + d)/r in comparison to that for the Mallard (*Anas platyrhynchos*) and presumably other dabblers that do not spread their wings. Since then, further studies on the wing-drying of cormorants have overwhelmingly supported the notion that its function is the drying of contour feathers and not thermoregulation, balancing, intraspecific signaling or an aid to swallowing fish [31]. Yet the relation between the cormorant's feather structure, specifically its parameter (r + d)/r, and this behavior has remained elusive. It is possible, though, to draw a number of conclusions from more recent data recorded with modern imaging software.

Water birds that regularly spread their wings include several species of cormorants, such as the Reed Cormorant (*P. africanus*), Bank Cormorant (*P. neglectus*), Cape Cormorant (*P. capensis*), White-breasted Cormorant and the Double-crested Cormorant, most of which have (r + d)/r values for their contour feathers between 4.3 and 4.9. Families and species with parameters under about 4.2, such as the divers (4.0), gannets (*Sulidae*) (3.8), auks (*Alcidae*) (3.4), penguins (2.3) and the Antarctic Blue-eyed Shag (*P. atriceps*) (3.8) never show wing-spreading behavior. Pelicans (*Pelecanidae*) (4.9 to 5.4), including the Brown Pelican (5.9), do so only very occasionally, but all other water birds do not with the notable exception of the darters (10 to 11). Darters have contour feathers that promote water to penetrate to the skin in order to reduce their buoyancy [34, 35] so their very large (r + d)/r value in unsurprising. It is reasonable to assume that, with the exception of the darters, all water birds benefit from a plumage with good water repellency and equally good resistance to water penetration. However, as we have seen, the structural requirements for these two qualities are partly opposed, so it is to be expected that each family or species will have made trade-offs and struck a balance that suits its specific demands of habitat and behavior. Spread-wing postures can then be explained as being part of a behavioral pattern in those birds that dive frequently and therefore require good resistance to water penetration, but this resistance comes at the expense of a measure of water repellency, which is compensated for by wing-drying.

The question as to whether it is the wings or the body plumage that is being dried by wing spreading was raised by Sellers [31] and can be addressed by the considering the difference in (r + d)/r values between flight and contour feathers. Values for flight feathers, in particular the outer coverts, measure 2 to 4 for both water and terrestrial birds, and these are therefore well protected from becoming waterlogged. Those for contour feathers, on the other hand, show much difference between these two groups of birds, with those of water birds that spend much time in the water and dive frequently ranging from about 2 to 4 and those of terrestrial birds ranging from about 7 to 10. Other water birds, including cormorants, have values that fall somewhere in between. Contour feathers with (r + d)/r values higher than about 4 are at risk of becoming waterlogged, which suggests that it is the exposed contour feathers rather than the flight feathers that need drying in cormorants and in darters.

Apart from (r + d)/r values, weather may also influence wing-spreading behavior. Cormorants reduce the extent to which they spread their wings with increasing wind speed, and at speeds of 4 on the Beaufort scale Sellers never saw birds to extend their wings by more than about 50%. Wind speeds may also be the reason why spread-wing postures are unknown in the Antarctic populations of the Blue-eyed Shag [29], but common in birds of this species breeding in Chile [36]. The persistent strong winds at high latitudes may well be the cause for the absence of wing-spreading behavior in the Antarctic populations.

Other than wind speed, the relative temperatures of water and air may be a factor in wing-spreading. A case in point is the Flightless Cormorant (*P. harrisi*) of the Galapagos, which is known to spread its stubby wings after a dive in the cool waters of the archipelago and, in this respect, behaves no different from other cormorants.

However, whereas most other cormorants have contour feather with barb diameters between 48 and 54 μm and (r + d)/r values between 4.3 and 4.9, those for the Flightless Cormorant are 31–41 μm and 7.1–7.4, respectively [21]. These numbers suggest that the Flightless Cormorant suffers a measure of water penetration through the barbs of its contour feathers, a feature that is more reminiscent of darters than of cormorants. As with darters, increased water penetration is thought to assist the underwater bottom-feeding habits of P. harrisi for which too much buoyancy would prove to be a disadvantage. Simple calculations appear to support this notion: the pressure that a surface-swimming Flightless Cormorant exerts on the water ranges between 630 and 780 N m^{-2}, whereas only 550 to 590 N m^{-2} pressure is required to force water between the barbs [37, 38]. For other cormorants, the maximum weight for no water penetration between the barbs lies well above the bird's weight range [26]. So, unlike those of other cormorants, the Flightless Cormorant's contour feathers become waterlogged after a dive in cold water, but the bird can then proceed to dry its plumage in the warm tropical breezes on the lava rocks, an advantage denied to cormorants inhabiting high latitudes.

6. Other behavioral patterns

In the previous section, we attributed the occurrence or absence of spread-wing postures to the need for a trade-off between water repellency and resistance as reflected in the value of the parameter (r + d)/r. It is therefore to be expected that other behavioral patterns, directly or indirectly, relate to this parameter in a similar manner. As an example of the relationship between 2r, (r + d)/r, and behavioral pattern, gannets, cormorants, and shearwaters (*Procellaridae*) all have about the same barb diameter (50–51 μm), but gannets have a value for (r + d)/r of 3.8, which lies at the low end of the range (3.8–4.9), indicating a greater resistance to water penetration. This may well be an adaptation to the gannet's habit of diving from the air and then pursuing prey under water. Brown Pelicans also dive from the air, but unlike gannets do not pursue their prey under water. Their breast feathers have smaller barb diameters and higher (r + d)/r values than those of gannets, producing an increased water repellency. American White Pelicans (*Pelecanus erythrorhynchos*), on the other hand, find their prey while swimming on the surface and have smaller values for (r + d)/r. Apparently, plungers, divers and birds that swim underwater benefit mostly from an increased resistance to water penetration, whereas surface feeders, such as the Brown Pelican, gulls and storm petrels (*Hydrobatidae*), profit from an increased water repellency. Similar findings were recorded for the five species of Dippers (*Cinclidae*), which among them show a slightly different water repellency and resistance in their contour feathers as an adaptation to their different feeding habits and river habitats [28]. Certain species of cranes (*Gruidae*) and rails (*Rallidae*) can also be regarded as having attained structural characters in their plumage that relate to their specific interaction with their watery feeding grounds [38–40].

It is likely that many more examples of contour feather structure correlating with specific behavior/habitat will be found once more data have been gathered. However, the above examples suffice to suggest that each feather substructure represents an evolutionary adaptation to a specific set of behavioral patterns and habitat conditions. It should be borne in mind that feather structure relates in the first place to behavior and habitat and secondarily to family identity and then only to the extent that family members behave in essentially the same way and inhabit similar habitats. As we have shown, congeners with different behavior/habitat patterns show a correspondingly different value for the structural parameter. That this behavior difference occurs in conjunction with a structural difference supports the existence of a correlation between feather structure and the habitat and behavior of its avian bearer.

7. Adaptations to the impact forces of diving, plunging and alighting

Unlike terrestrial birds, water birds physically interact with water at the interface between feather coat and water. Water is about 800 times denser than air and, as a result, the impact forces of diving, plunging and alighting are so much more severe than when operating in air. It is no surprise therefore, that the feathers of water bird are composed of stiffer elements to cope with these conditions. Since each family interacts with water in its own specific way variations in feather stiffness among families are to be expected for this reason alone.

All feathers are built of beta-keratin the elastic modulus of which is an inherent property of the keratin material itself. However, the actual stiffness of the various feather elements, rachis, barbs and barbules, is determined by the respective shapes and sizes of these elements.

The mechanical forces involved in diving, plunging and alighting are not accessible to direct measurement in any reliable or representative way, at least not by current technologies. Any such data would not be meaningfully correlated to the resulting yield or flexure of barbs and vanes during forceful interaction with water. However, the bending and flexing of materials of different shapes and sizes have been well described in engineering physics and it is from these considerations that a number of conclusions can be drawn [41]. For one, the bending of the vane as a result of a force acting over its surface can be semi-quantitatively determined from the length, width and separation of the barbs, all of which can be easily measured. Specifically, the ratio of length to width of a barb can be shown to be the predominate factor in resisting bending of the vane, more so than the other contributor $(r + d)/r$. The deflection parameter (DF) that quantifies the bending per unit force applied to the vane is the product of this ratio and the wettability parameter $(r + d)/r$. For another, apart from feather stiffness, the resistance to impact forces is also determined by the extent of contour feather overlap and body feather density. The extent of overlapping can be approximated by the product of L_f, the length of the rachis, and the square root of the number of feathers per surface area. To estimate the latter, use was made of the data on number of feathers and body weights as reported by several authors [41–47]. By fitting a second-order polynomial to these data, an estimate of the number of contour feathers as a function of the mass of the bird could be obtained. For the relationship between body surface area and body mass, expressions proposed by Perez, Moye and Pritsos [48] and by Mitchell [49] were used to estimate surface area as a function of body weight. Combining the results of these two sets of calculations, contour feather densities expressed in number of feathers per surface area were found to be about 100,000 to 150,000 per m^2 for water birds weighing less than 1.2 kg for all families studied. This number increases with weight to 200,000/m^2 at about 7 kg. The extent of feather overlap, according to these calculations, yields about 10 to 15 feathers in a stack for families in the lower weight range with twice that number for heavier birds. Apparently, feather overlapping is the same for water birds in the lower weight range regardless of family identity and, as a result, the restriction that stacking provides to bending is also the same. Only for birds weighing more than 1.2 kg do we find an increase in feather density and overlap with weight – up to 250,000 per m^2 and stacks of 18 for a Pink-backed Pelican (*P. rufescens*) weighing 9.6 kg. This is in line with expectation as impact forces are directly proportional to mass [50].

The role of the barbules in resisting bending of the vane should be considered in the light of their primary function, i.e., keeping the barbs from separating under an applied force and doing so by their hooks sliding in the flanges of the barbule next more distal. Therefore, as well as for their small size, they are assumed to make only a minimal, if any, contribution to the over-all resistance to bending.

The above findings may be explained by any of two or both possibilities: 1) the feather density and number of feathers in a stack for the lower-weight families are sufficiently large to prevent feather bending regardless of behavioral pattern and 2) barb stiffness and resistance to water penetration of the contour feathers of each of these families are large enough to prevent water from reaching the skin on their own account and do not benefit from a further increase in feather density or stacking. Other than preventing water from reaching the skin, thermoregulatory adaptations can also be expected to affect feather density. Lowe [46] counted 48/cm^2 on a young Gentoo Penguin (*Pygoscelis papua*).

As mentioned earlier, the bending of the vane of a contour feather under the impact of forces associated with diving or landing on water surfaces consists of two factors: (1) the ratio of the length to the thickness of the barbs and (2) the wettability parameter $(r + d)/r$. The first factor indicates that short and thick barbs make the vane stiff resisting bending, whereas long and thin barbs favor flexibility that promotes bending. The appearance of the wettability parameter in the equation shows that feathers resistant to water penetration also help prevent their bending, whereas highly water-repellent feathers do not.

The deflection parameter values for 23 water bird species from 15 families have been assorted into six more or less distinct categories [50]. Deep divers, represented by four species of penguins and characterized by their habit of diving and pursuing prey under water, fall in the lowest category with a DF of 1.6 (10^6) and therefore have the highest vane stiffness. The next category is made up of birds that swim and dive in pursuit of their prey and spend much time in and on the water. This category has a DF of 49 (10^6) and includes the Common Diving Petrel (*Pelecanoididae*) and the cormorants. Category 3, into which fit the 'true plungers' such as petrels (*Procellariidae*), gannets and auks, show a DF of 194 (10^6). Large surface feeders, such as pelicans, frigatebirds (*Fregatidae*) and skimmers (*Rhynchopidae*), form the next category with a DF of 387 (10^6). Category 5, the shore birds, includes skuas (*Stercorariidae*), gulls and terns (*Sternidae*) that have the lowest vane stiffness with a DF of 839 (10^6). These birds are not extended time swimmers, do not pursue their prey under water and spend much time in flight or on shore. Albatrosses, ranked as large birds of open oceans, are mostly airborne and alight only to take food from the surface or slightly below. In this respect, they behave much like category 5 families. Not listed are the Flightless Cormorant in category 2 and the Brown Pelican in category 4, because, as mentioned above, these species feed in a different way from their congeners, a feature expressed in the dimensions of their feather structure.

From these data can be concluded that the contour feathers of penguins, the most aquatic of families, are about thirty times stiffer than those of diving petrels and cormorants, and 120 times more so than those of plungers like gannets. Similarly, penguin feathers are 250 times more resistant to bending than those of surface feeders like pelicans, over 500 times more so than those of shorebirds such as skuas, gulls and terns and almost three orders of magnitude stiffer than albatross feathers. These large differences are directly related to feeding habits and interaction with water. Penguins find their prey exclusively under water and dive to great depths to catch it. Diving petrels and cormorants also dive, but spend more time on the surface and in the air. Plungers dive from the air with associated high pressure on impact, but catch their prey at lesser depths. Surface feeders do not dive and do not pursue their prey under water (brown pelicans dive from the air, but do not pursue under water). Shore birds feed from the water surface and are not extended time swimmers. Albatrosses, one of the most aerial of seabirds, alight only to feed from the surface and may occasionally dive at feeding frenzies.

The following pattern of feather structure in relation to feeding habits/behavior emerges:

Barb width and spacing determine the relative water repellency and resistance to water penetration of feathers. Diving birds, and in particular deep diving birds, benefit from a mostly water resistant plumage with little in the way of water repellency. Less aquatic families, such as gannets and to a greater extent cormorants, show more repellency, but at the expense of some of their water resistance. Some cormorants compensate for this by their habit of wing spreading. Swimming and hovering birds that catch their prey from the surface, shore birds and those operating mostly in the skies show a predominantly water repellent plumage.

The length and diameter of the barbs of contour feathers vary widely among water birds. Barb stiffness varies with barb length and width and is largest for deep diving birds, less so plungers and very much less so for surface feeders ranging over three orders of magnitude. These structural differences in the feather plumage are thought to represent evolutionary adaptations to feeding habits and, in some cases, environmental conditions.

8. Viscous drag and feather geometry

Birds that swim and dive will also benefit from reduced drag for their locomotion in water, a consideration that applies less to waders and shore birds and not at all to land birds. Viscous drag in water is dependent on the surface microstructure of the distal one-third of the contour feather, but also on the shape of its surface in contact with water, an aspect of feathers that has so far received little or no attention. Drag in air, such as in flight, on the other hand, has been the topic of several studies.

That the shape of the surface area in contact with water varies among bird families has been noticed in the course of previous studies. It was seen to be nearly circular in land birds with a length-to-width ratio (L/W) of approximately 1.0, but oblong with an L/W of about 4 in penguins, the most aquatic of families. Birds less intimate with open water showed intermediate values for L/W.

Here, we consider the interface between the distal one-third and flowing water to calculate viscous drag for feather shape geometry. Assuming the flow to be parallel to the long axis of the feather, i. e. zero angle of attack, one can derive the total drag coefficient, composed of viscous pressure and frictional drag, from the computational and experimental results of studies on model ship hulls of varying length-to-diameter ratios using solutions to the Reynolds-averaged Navier-Stokes equations [51]. The drag coefficient then follows from the value of L/W for values of L/W less than 7 which is within the range of feather geometry. The relationship between drag coefficient and L/W predicts that oblong shapes of the tips of contour feathers reduce drag facilitating swimming and diving, whereas a more circular shape would cause an increase in frictional drag. A similar reasoning could be applied to the shape of the area that the body of a swimming bird has in contact with water. If this area is assumed to be elliptical, a drag coefficient for body surface area in contact with water can be determined using the same considerations.

Grouping bird species according to their interaction with open water can be achieved by assigning them to foraging niches as proposed by Pigot et al. [52], applying a standardized protocol for foraging niche delimitation. Using this procedure, a total of thirty niches has been identified for all of the approximately 10,000 bird species of the world. Of these six major foraging niches were categorized as Aquatic.

Aquatic birds can be subdivided into swimmers and waders. Their values for the drag coefficient show a viscous drag for swimmers significantly lower than that of waders and, predictably, land birds. These categories, when further subdivided into eight aquatic foraging niches and, for comparison, two terrestrial ones, show that divers have the lowest recorded drag coefficient increasing in order for plungers, surface feeders, aerials, herbivore surface feeders, ground feeders, perchers to herbivore ground feeders. Land birds experience an even higher drag with no significant difference between ground feeders and those that catch their prey by aerial or sally sorties.

Adaptations in feather microstructure and body surface area in contact with water that bring about a reduction in viscous and frictional drag while swimming increase according to the bird's intimacy with open water. Swimming and diving birds, such as penguins and grebes, benefit the most from reduced viscous drag, more so than plungers such as gannets. Aerials such as terns even less so, but much more than herbivore surface feeders such as ducks. The body feathers of ducks, in turn, appear to be better adapted to their watery habitat than those of aquatic ground feeders such as herons or kingfishers. The drag coefficient of the Spotted Dikkop (*Burhinus capensis*), an herbivore aquatic ground feeder of open scrubby habitat with comparatively little interaction with open water, was found to be more in line with those of land birds in which adaptations to locomotion in water are not expected to have evolved.

Land birds do not only show drag coefficients higher than those of water birds, they also show no significant difference among the two foraging niches examined so far. This is in line with expectation as their lack of interaction with open water and their locomotion in air only suggest that forces that foster reduced drag in water have been absent in their evolutionary history.

Author details

Arie M. Rijke
University of Virginia, USA

*Address all correspondence to: amr@virginia.edu

References

[1] Landsborough Thomson A. A New Dictionary of Birds. London: Nelson; 1964

[2] Stettenheim PR. The integument of birds. In: Farner DS, King JR, editors. Avian Biology. Vol. vol. 2. New York: Academic Press; 1972. pp. 1-63

[3] King JR, Farner DS. Energy metabolism, thermoregulation and body temperature. In: Marchall AJ, editor. Biology and Comparative Physiology of Birds. New York and London: Academic Press; 1961. pp. 215-288

[4] Lucas AM, Stettenheim PR. Avian Anatomy. Integument. Agricultural Handbook 362. Washington, DC: US Dept. Agr; 1972

[5] Dove CJ. Quantification of microscopic feather characters used in the identification of North American Plovers. Condor. 1997;**99**:47-57

[6] Dove CJ. A descriptive and phylogenetic analysis of plumulaceous feather characters in Charadriiformes. American Ornithologists Union, Ornithological Monograph. 2000;**2000**:51

[7] Doucet SM, Shawkey MD, Hill GE, Montgomery R. Iridescent plumage in satin bower birds: structure, mechanisms and nano-structural predictors of individual variation in color. Journal of Experimental Biology. 2006;**209**:380-390

[8] Stettenheim PR. The integumentary morphology of modern birds – an overview. American Zoology. 2000;**40**:461-477

[9] Lei FM, Qu YH, Gan YL, Kaiser M. The feather microstructure of passerine sparrows in China. Journal of Ornithology. 2002;**143**:205-213

[10] Dove CJ, Rijke AM, Wang X, Andrews LS. Infrared analysis of contour feathers. The conservation of body heat radiation in birds. Journal of Thermal Biology. 2007;**32**:42-46

[11] Christiansen C. Untersuchungen uber die optischen eigenschaften von fein vertheilten korpern. Annalen der Physik und Chemie. 1884;**23**:298-306

[12] Chistiansen C. Untersuchungen uber die optischen eigenschaften von fein vertheilten korpern. Annalen der Physic und Chemie. 1885;**24**:439-446

[13] Ljungdahl G, Ribbing CG. Infrared properties of diabase glass and wool. Journal of Applied Physics. 1989;**65**(12):5024-5030

[14] Cassie ABD, Baxter S. Wettability of porous surfaces. Transactions of the Faraday Society. 1944;**40**:546-551

[15] Rijke AM, Jesser WA. The water penetration and repellency of feathers revisited. The Condor. 2011;**133**:245-254

[16] Moilliet JL. Water Proofing and Water Repellency. New York: Elsevier; 1963

[17] Rijke AM. The liquid repellency of a number of fluoro-chemical finished cotton fabrics. Journal of Colloid Science. 1965;**20**:205-216

[18] Bormashenko E, Bormashenko Y, Stein T, Whyman G, Bormashenko E. Why do pigeon feathers repel water? Hydrophobicity of pennae, Cassie-Baxter wetting hypothesis and Cassie-Wenzel capillarity-induced wetting transition.

Journal of Colloid and Interface Science. 2007;**311**(1):212-216

[19] Ma M, Hill RM. Superhydrophobic surfaces. Current Opinion in Colloid & Interface Science. 2006;**11**:193-202

[20] Adam NK, Elliot GEP. Contact angles of water against saturated hydrocarbons. Journal of the Chemical Society. 1962;**424**:2206-2209

[21] Shafrin EG, Zisman WA. The spreading of liquids on low-energy surfaces IV: monolayer coatings on platinum. Journal of Colloid Science. 1952;7:166-177

[22] Shafrin EG, Zisman WA. The adsorption on platinum and wettability of monolayers of terminally fluorinated octadecyl derivatives. Journal of Physical Chemistry. 1957;**61**:1046-1053

[23] Rijke AM, Jesser WA, Mahoney SA. Plumage wettability of the African Darter (*Anhinga melanogaster*) compared with the Double-crested Cormorant (*P. auritus*). Ostrich. 1989;**60**:128-132

[24] Elder WH. The oil gland of birds. Wilson Bulletin. 1954;**66**:6-31

[25] Odham G, Stenhagen E. On the chemistry of preen gland waxes of water fowl. Accounts of Chemical Research. 1971;**4**:21-128

[26] Rijke AM. Wettability and phylogenetic development of feather structure in water birds. Journal of Experimental Biology. 1970;**52**:469-479

[27] Rijke AM, Burger EH. Wettability of feathers and behavioral patterns in water birds. Proceedings of the Pan-African Ornithological Congress. 1985;**6**:153-158

[28] Rijke AM, Jesser WA. The feather structure of dippers: water repellency and resistance to water penetration. Wilson Journal of Ornithology. 2010;**122**:563-568

[29] Bernstein NP, Mason SJ. Absence of wing-spreading behavior in the Antarctic Blue-eyed Shag (*Phalacrocorax atriceps bransfieldensis*). Auk. 1982;**99**:588-589

[30] Elowson AM. Spread-wing postures and the water repellency of feathers: a test of Rijke's hypothesis. Auk. 1984;**101**:371-383

[31] Sellers RM. Wing-spreading behavior of the cormorant, *P. carbo*. Ardea. 1995;**83**:27-36

[32] Mahoney SA. Plumage wettability of aquatic birds. Auk. 1984;**101**:181-185

[33] Rijke AM. The water repellency and feather structure of cormorants, phalacrocoracidae. Journal of Experimental Biology. 1968;**48**:185-189

[34] Owry OT. Adaptations for locomotion and feeding in the Anhinga and Double-crested Cormorant. Ornithological Monographs. 1967;**6**:60-63

[35] Mahoney SA. Some aspects of the thermal physiology of Anhingas (*Anhinga anhinga*) and Double-crested Cormorants (*P. auritus*). In: Cooper J, editor. Proceedings of the symposium on birds of the sea and shore, 1979. Cape Town: African Seabird Group; 1981. pp. 461-470

[36] Rasmussen P, Humphrey PS. Wing-spreading in Chilean Blue-eyed Shags (*P. atriceps*). Wilson Bulletin. 1988;**100**:140-144

[37] Brudis J, Bass F. Report to the Charles Darwin Research Station, Santa Cruz, Galapagos. Punta Espinosa, Fernandina Island: Field Work; 1979

[38] Rijke AM. Wettability and Feather Structure of Endemic Galapagos Water Birds. Puerto Ayora, Santa Cruz Island, Galapagos, Ecuador: Charles Darwin Research Station; 1986

[39] Tarboton WR. The Complete Book of Southern African Birds. Cape Town: Struik Winchester; 1990

[40] Rijke AM, Jesser WA. Plumage wettability of African cranes. Proceedings of the Pan-African Ornithological Congress. 1993;**8**:533-537

[41] Wetmore A. The number of contour feathers in passeriform and related birds. The Auk. 1936;**53**:159-169

[42] Hutt FB, Ball L. Number of feathers and body size in passerine birds. The Auk. 1938;**55**:651-657

[43] Dwight J. The sequence of plumages and molts of the passerine birds of New York. Annals of New York Academic Science. 1900;**13**:118-119

[44] McGregor RC. The number of contour feathers in passeriform and related birds. The Auk. 1903, 1936;**53**:159-169

[45] Knappen P. Number of feathers on a duck. The Auk. 1932;**49**:461

[46] Lowe PR. The number of contour feathers in passeriform and related birds. The Auk. 1933;**53**:159-169

[47] Kuhn O, Hesse R. Die postembryonale Pterylose bei Taubenrassen verschiedener grosse. Z Morphol Oekol Tiere. 1957;**45**:616-655

[48] Perez CR, Moye JK, Pritsos CA. Estimating the surface area of birds: using the homing pigeon (*Columba livia*) as a model. Biological Open. 2014;**3**:486-488

[49] Mitchell HH. The surface area of single comb white leghorn chickens. Journal of Nutrition. 1930;**2**:443-449

[50] Rijke AM, Jesser WA, Barnard GR, Coertze RD, Bouwman H. The contour feathers of water birds exhibit adaptations to the impact forces of diving, plunging and alighting. Ecology and Conservation Science Open Access. Feb 2023;**2**(3). DOI: 10.19080/ECOA.2023.02.555585

[51] Lin CW, Percival P, Gotimer EH. Viscous drag calculations for ship hull geometry. In: Ninth International Conference on Numerical Methods in Laminar and Turbulent Flow. Atlanta; 1995

[52] Pigot AL, Sheard C, Miller ET, Bregman TP, Freeman BG, Roll U, et al. Macroevolutionary convergence connects morphological form to ecological function in birds. National Ecological Evolution. 2020. DOI: 10.1038/s41559-019-1070-4

Chapter 8

Histological Analysis of Different Organ Systems in Ostrich

Julio Luis Erwin Illanes Herrero

Abstract

In the last decade, ostrich breeding in Chile has increased substantially, so it is necessary to increase studies to improve and understand the adaptation of this species to our ecology. Our objective is to contribute to the knowledge of the normal histology of the glands attached to the digestive system: liver, proventriculus and pancreas; digestive system: esophagus, proventriculus, muscular stomach, small intestine (duodenum, jejunum and ileum), large intestine (cecum, colon and rectum); respiratory system: larynx, trachea, syrinx, extrapulmonary primary bronchi, intrapulmonary primary bronchi or mesobronchi, secondary bronchi, parabronchi or tertiary bronchi, air capillaries and air sacs; and female reproductive system: ovary and oviduct (infundibulum, magnum, isthmus, uterus, and vagina) of *Struthio camelus var.* domesticus in a first stage. This research was carried out with clinically healthy ostriches, from which samples of the different systems were obtained for this study. Histological sections were made, which were stained and mounted for their comparative analysis under a light microscope, between ostrich and chicken. The comparative histological analysis allowed us to establish that the structure of the different apparatuses and systems is similar to that observed in the chicken. However, at the microscopic level some differences are observed.

Keywords: ostrich, histology, digestive system glands, digestive system, respiratory system, female reproductive system

1. Introduction

There are currently four subspecies of ostrich: *Struthio camelus camelus*, *Struthio camelus molybdophanes*, *Struthio camelus massaicus* and *Struthio camelus australis* [1]. In order to obtain an animal with greater productivity and easier handling, the domestic ostrich, *Struthio camelus* var. domesticus, result from the cross between two subspecies, *Struthio camelus camelus* and *Struthio camelus massaicus* [2]. The main products obtained from this bird are meat, skin, feathers, and eggs [1]. Histology in this species has not been extensively studied, which has prevented, in a certain way, knowing and satisfying its nutritional requirements with greater certainty [3]. Due to the increase observed in the use of the ostrich in the industry, its study acquires interest to obtain the necessary knowledge that allows to prevent, treat, and cure; as well as maintaining optimal preservation of this species [1].

The digestive system presents glands such as the proventriculus (glandular stomach), liver, and pancreas that have been anatomically described. The true stomach of the ostrich corresponds to the proventriculus or glandular stomach, it is sac-shaped and anatomically occupies the cranial part of the abdomen in the left hypochondrium [4], where protein digestion begins through the secretion of pepsinogen and hydrochloric acid [3]. In relation to the general anatomical characteristics of the digestive tract, it is known that the ostrich presents similarities and differences, both with other birds, as well as with ruminants and other herbivores (equines) [3]. The comparative analysis with the hen presents certain anatomical differences, one of them is not having a crop, a food storage organ that exists in other birds. The proventriculus and the muscular stomach (gizzard), in the ostrich, can fulfill this function [5]. They do not have a gallbladder, so the emptying of bile is done directly into the small intestine. In addition, the large intestine of the ostrich, unlike other species, represents 50% of the total length of the digestive tract and the small intestine corresponds to only 35.5% [2].

Among the anatomical characteristics of the ostrich respiratory system, it is pointed out that the larynx is formed by two arytenoid cartilages and one cricoid cartilage, which together form a ring [1]. The trachea extends between the larynx and the syrinx, with the dorso-ventrally compressed lumen in the upper third. It is made up of approximately 200 complete cartilaginous or osseous rings [1]. The syrinx is simple, formed by the last rings of the trachea and the proximal half rings of the extrapulmonary bronchi. These extend into the lungs only a short distance [1]. Regarding the lungs, they are located in the dorsal third of the thoracoabdominal cavity, between the second and seventh pair of ribs [1]. Like flying birds, the ostrich has air sacs incorporated into its anatomical structure, in a number of five pairs (clavicular, interclavicular, pre-thoracic, post-thoracic, and abdominal). In birds, many of the bones have cavities that communicate with the air sacs, which is why they are called pneumatic bones. In the case of the ostrich (*Struthio camelus*), the only pneumatic bones present are the sternum and the femur [6].

The reproductive organs of the female ostrich are the ovary and left oviduct that empties into the urodeum, with the rudimentary right ovary being [1, 6]. In the ostrich, the ovary is suspended from the dorsal body wall, ventral to the kidney and dorsomedial to the ipsilateral abdominal air sac. The position, size, and shape of the ovary depend on seasonality and vary during the breeding season [2]. The oviduct carries the fertilized oocyte to the cloaca, and in the opposite direction allows sperm to reach the fertilization site [7]. The oviduct of birds has five regions each with their respective function, these are infundibulum, magnum, isthmus, uterus and vagina [8]. The infundibulum receives and encloses the yolk, and it is the site where fertilization occurs [9]. The albumin or white is secreted in the magnum and is the longest segment of the oviduct. In the isthmus, the testaceous membranes of the shell form around the egg [2]. Calcification of the shell occurs in the uterus (or shell gland) which is globose in shape. The vagina in birds allows the passage of the egg formed from the uterus to the cloaca during posture or oviposition and is also important in the selection, transport, and storage of spermatozoa [10]. The ostrich reaches sexual maturity in the third year of life, but with nutritional management, it can be brought forward to the second year of life. The ostrich has a seasonal breeding period of long days and can lay up to 50 eggs per season. The time it takes between the formation and laying of an egg is 48 hours compared to the hen which is approximately 24 hours [2].

The objective of this work is to identify, classify, and histologically characterize the glands associated with the digestive system: liver, proventriculus, and pancreas; the different segments of the digestive system: esophagus, proventriculus, muscular stomach, small intestine (duodenum, jejunum, and ileum), and large intestine (cecum, colon, and rectum); respiratory system: larynx, trachea, syrinx, extrapulmonary primary bronchi, intrapulmonary primary bronchi or mesobronchi, secondary bronchi, parabronchi or tertiary bronchi, air capillaries, and air sacs; and female genital apparatus: ovary and oviduct comprising: infundibulum, magnum, isthmus, uterus, and vagina of *Struthio camelus* var. domesticus. These were comparatively analyzed with what was described in the hen (*Gallus gallus*), with those species that are morphologically associated with them, such as domestic birds, ruminants, pseudo-ruminants (camelids), and eventually other species.

2. Material and methods

The biological material used in this work corresponds to samples of glands associated with the digestive system and the different segments of the digestive, respiratory, and female reproductive systems, obtained from clinically healthy adult ostriches (*Struthio camelus* var. domesticus), with a range from 12 to 14 months of age, from farms in the Sixth Region of Chile. The material was obtained from the FAENAGRO and CAREN fridge located in VI Region, Chile. Once the samples were obtained from the different systems, they were fixed in 10% buffered formalin and then followed with the usual histological technique procedure. The stains used were the current hematoxylin-eosin (HE) techniques; Mallory's trichrome, and Van Gieson's, in order to observe muscle and connective tissues [11]. Sections 5 μm thick were mounted, stained, photographed, and analyzed under a light microscope (Olympus BX41 with Olympus C-5060 Wide zoom digital camera, 5.1 megapixels). With the photographs, a comparative analysis of the histological aspects found in ostrich, with those described in other species, was carried out.

3. Glands attached to the digestive system

The glands attached to the ostrich digestive system correspond to: liver, which has 4 lobes and is located in the right hypochondrium; proventriculus, which presents a wide lumen and 4 layers recognized in its wall: mucosa, submucosa, muscular and serous; and pancreas, it is a mixed gland, exocrine and endocrine, very lobed, made up of parenchyma (pancreatic islets and glandular acini), and a stroma (capsule, septa, and hilum), [12].

Liver: The parenchyma is irregularly septated and made up of cords of hepatocytes arranged radiating from a central vein. Hepatocytes generally have one or two nuclei. Sinusoidal capillaries were found between the cords, where macrophages were observed (**Figures 1** and **2**). It should be noted that the hepatocytes located around the portal spaces are also arranged radially (**Figure 3**). From the capsule arise the partitions that divide the organ irregularly. The portal space corresponds to the loose connective tissue that contains blood vessels derived from the portal vein and the hepatic artery, lymphatic vessels, and bile ducts, the latter presenting simple cuboidal lining epithelium (**Figures 1, 3**, and **4**).

Proventriculus (Mucosa and Submucosa): The mucosa with simple prismatic epithelium is covered with mucus (glycocalyx). With a large number of villi and between its bases crypts with simple cuboidal epithelium are observed. The lamina propria of loose connective tissue with simple or branched tubular glands secreting

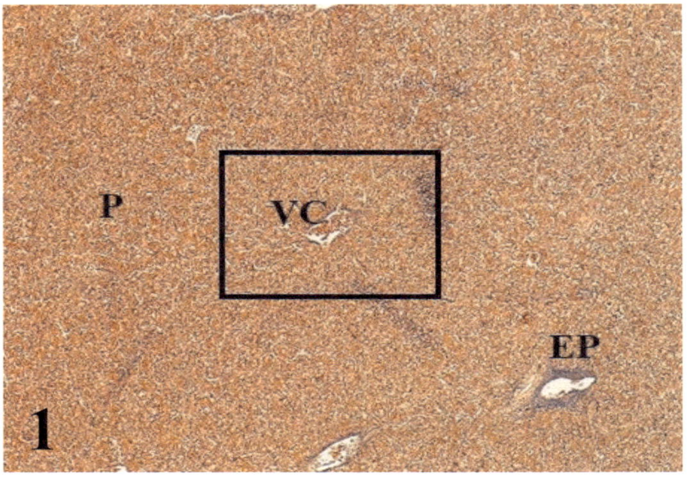

Figure 1.
Histological section of a liver lobule. VC: central vein; P: parenchyma and EP: portal space. Mallory's trichrome. 100×.

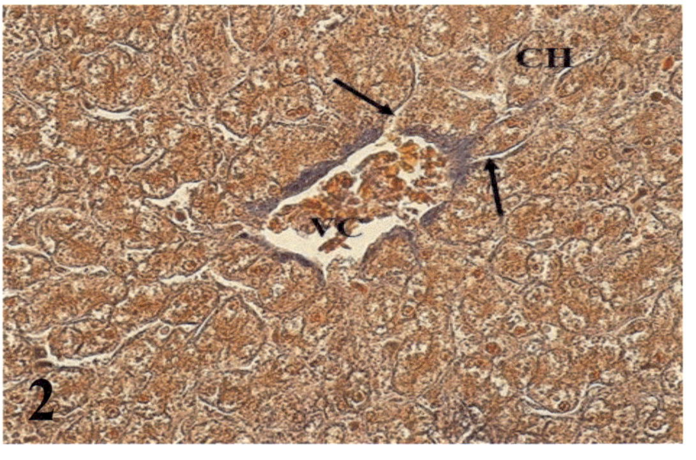

Figure 2.
Higher magnification of the inset of ***Figure 1****. VC: central vein of the lobule; CH: cords of hepatocytes and sinusoidal capillaries that drain into the central vein (arrows). Mallory's trichrome. 400×.*

pepsinogen and hydrochloric acid, extensive lymphocyte infiltration (**Figures 5** and **6**). The muscularis mucosa of smooth muscle tissue, not well defined, emits septa into the submucosa. The submucosa is thick and made up of dense connective tissue that emits septa around the compound tubuloalveolar glands, constituting lobules formed by conglomerates of tubuloalveolar adenomeres around a central lumen (**Figures 5** and **6**). The glands discharge their secretion into the lumen of the proventriculus, through its excretory duct lined by simple columnar epithelium. Among the cubic and cylindrical secretory cells of the adenomeres, the parietal cells, which produce pepsinogen and hydrochloric acid, are abundant. Ganglion cells of the submucosal plexuses are also seen in the submucosa of the proventriculus.

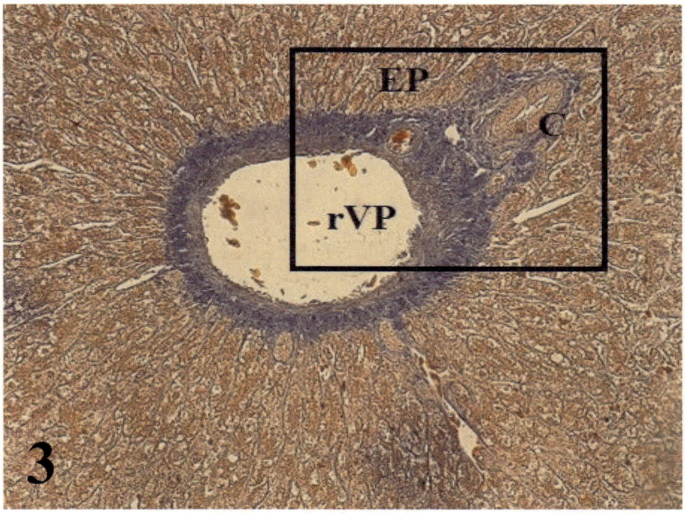

Figure 3.
Histological section of a portal space. EP: portal space; rVP: branch of the portal vein and C: bile duct. Note the ray arrangement of the hepatocytes around the portal space. Mallory's trichrome. 200×.

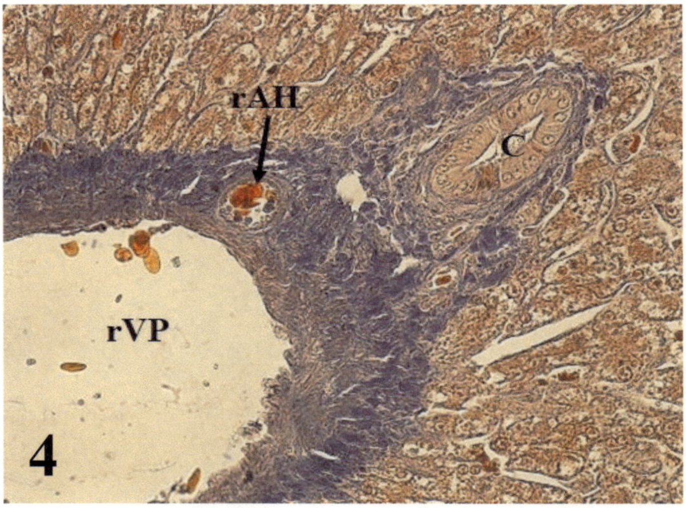

Figure 4.
Higher magnification of the inset of ***Figure*** *3. rVP: portal vein branch; rAH: branch of the hepatic artery and bile duct (C). Mallory's trichrome. 400×.*

Pancreas: The exocrine parenchyma is formed by tubuloacinar adenomeres composed of serous secretion and highly irrigated cell groups, which correspond to pancreatic islets that fulfill the endocrine function. The cells that constitute the adenomeres are prismatic with a round nucleus located in the basal third. The staining affinity of this zone is basophilic and the apical one is acidophilic (**Figures** 7 and 8). In the lumen of the acini, centro-acinous cells are observed that correspond to the beginning of the excretory ducts. The stroma is made up of the capsule that surrounds the organ, the septa, and the hilum. The capsule is relatively thin, made up of loose connective tissue that forms partitions toward the interior of the parenchyma, dividing it into small lobes and lobules.

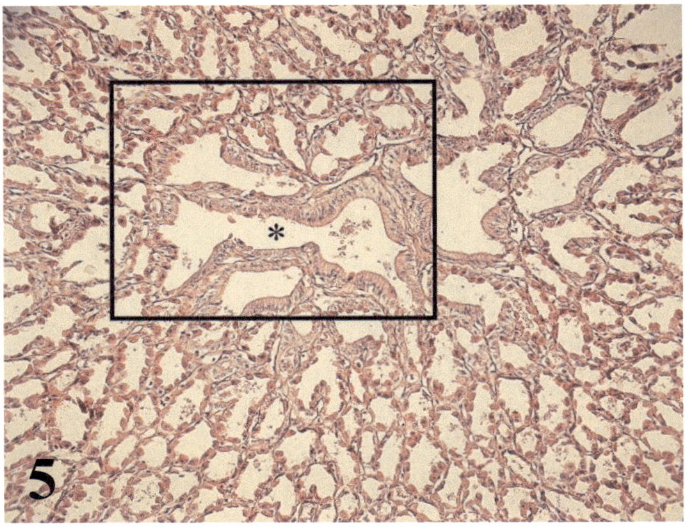

Figure 5.
Histological section of the submucosa of the proventriculus. (*): Excretory duct and tubuloalveolar adenomeres of the compound submucosal glands that empty into it. Staining: H-E. 200×.

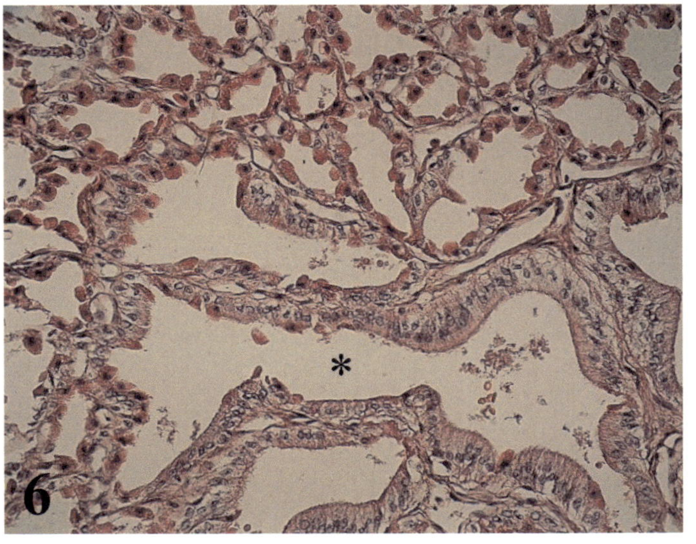

Figure 6.
Higher magnification of the inset of **Figure** *5. (*): Excretory duct of the submucosal glands and tubuloalveolar adenomeres with parietal cells (stained red). Staining: H-E. 400×.*

4. Digestive system

The segments that make up the digestive system correspond to the esophagus, proventriculus (glandular stomach), muscular stomach (ventricle or gizzard), small intestine, and large intestine. The wall of the digestive system presents four concentric layers or tunics: mucosa; submucosa, muscular and serosa, and/or adventitia. Throughout the digestive system, there are histological differences in each layer,

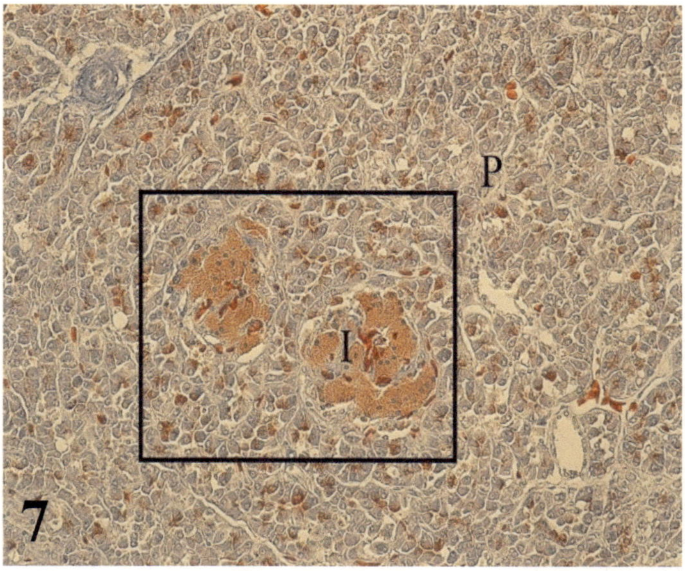

Figure 7.
Histological section of the pancreatic parenchyma. I: pancreatic islets and P: parenchyma. Mallory's trichrome. 200×.

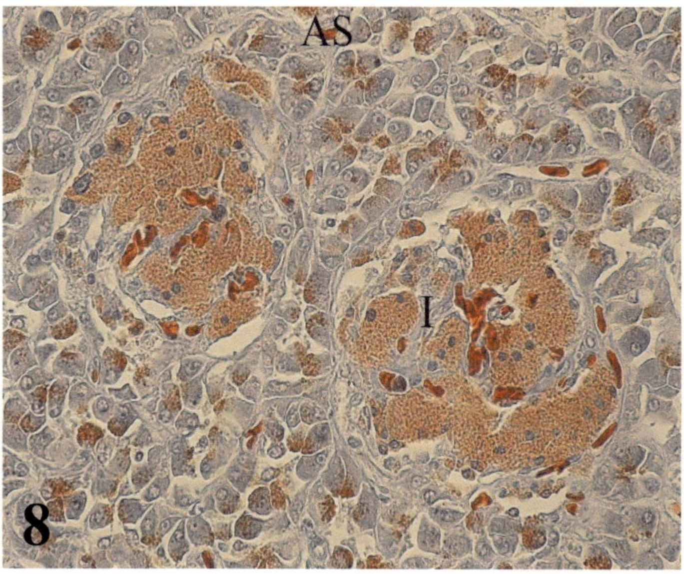

Figure 8.
Higher magnification of the inset of ***Figure 7****. I: pancreatic islets and AS: serous acinus. Note the blood supply to the pancreatic islet. Mallory's trichrome. 400×.*

mainly in the tunica mucosa, which are manifested by the functional characteristics of each segment [13].

Esophagus: A mucosa with a non-keratinized stratified squamous epithelium. A lamina propria or chorion of dense connective tissue with abundant simple and branched tubuloalveolar mucous secretion glands throughout its length. A large

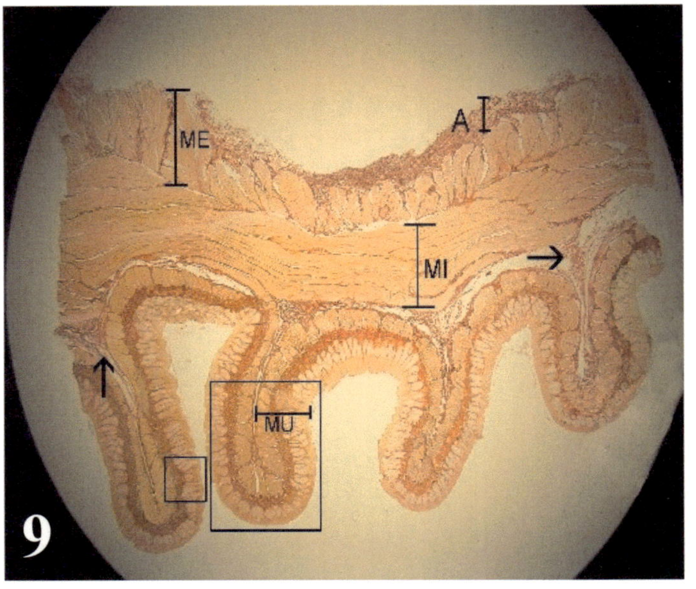

Figure 9.
Panoramic cross-section of the esophagus. MU. mucous membrane; submucosa (arrow); MI: internal muscle; ME: external muscle; A: adventitia. Note the longitudinal folds. Staining: Van Gieson 12.5×.

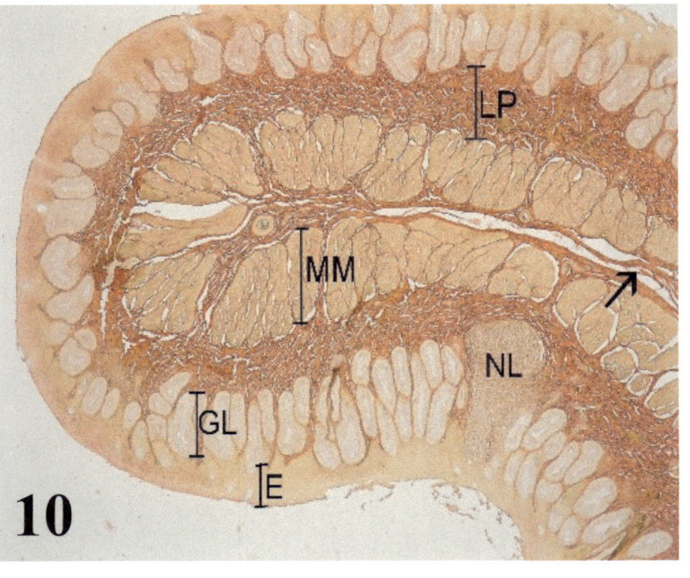

Figure 10.
*Higher magnification of the large box in **Figure 9**. E: nonkeratinized stratified squamous epithelium; GL: simple tubuloalveolar glands; LP: lamina propria; MM: muscular mucosa; NL: lymph node and submucosa (arrow). Staining: Van Gieson 100×.*

number of lymph nodes are observed. The muscularis mucosa presents only a thick layer of longitudinal smooth muscle fibers and determines the boundary between the lamina propria and the submucosa (**Figures 9** and **10**). The submucosa is poorly developed, made up of loose connective tissue with submucosal nerve plexuses along its structure. The muscularis is made up of a thick internal circular layer of smooth muscle and a more irregular and thin external longitudinal layer. Between both muscle layers, myenteric nerve plexuses can be seen. The serosa is made up of the mesothelium and loose connective tissue.

Proventriculus (glandular stomach): The mucosa with the simple prismatic epithelium of mucus secretion, with numerous villi and covered by a large amount of mucus (glycocalyx). The simple and branched tubular fundic glands, with principal and parietal cells, empty their secretion into the crypts. A lamina propria of loose connective tissue. A muscularis mucosa of smooth muscle tissue. Compound tubuloalveolar-type glands with parietal cells and excretory ducts of simple prismatic epithelium are observed in the submucosa. Submucosal plexuses are observed. The muscularis formed by a thick internal circular layer and a thin external longitudinal layer of smooth muscle. Between both layers myenteric plexuses are observed. The serosa is constituted by the mesothelium and loose connective tissue. Between the muscular and serous layers, we find a thick sheet of compact connective tissue with parallel bundles; this tendon layer surrounds the entire organ (**Figures 11** and **12**).

Muscular stomach (gizzard): The mucosa with simple prismatic epithelium is covered by a thick horny-looking layer, which corresponds to the cuticle (glycocalyx). It has a large number of wrinkles on its surface. A lamina propria of loose connective tissue with numerous simple and branching tubuloalveolar glands of mucous secretion. The muscularis mucosa is seen as a thin layer of smooth muscle fibers. These muscle fibers run parallel to the longitudinal axis and, in some portions, merge with the muscularis tunica. The submucosa of irregular dense connective tissue with septa branching into the muscular layer (**Figure 13**). The muscularis is thick, showing three layers of smooth muscle that correspond to an inner plexiform layer, a middle transverse layer, and a thick outer longitudinal layer (**Figure 13**). Numerous myenteric plexuses are observed between the muscle layers (**Figure 14**). The serosa is constituted by the mesothelium and loose connective tissue.

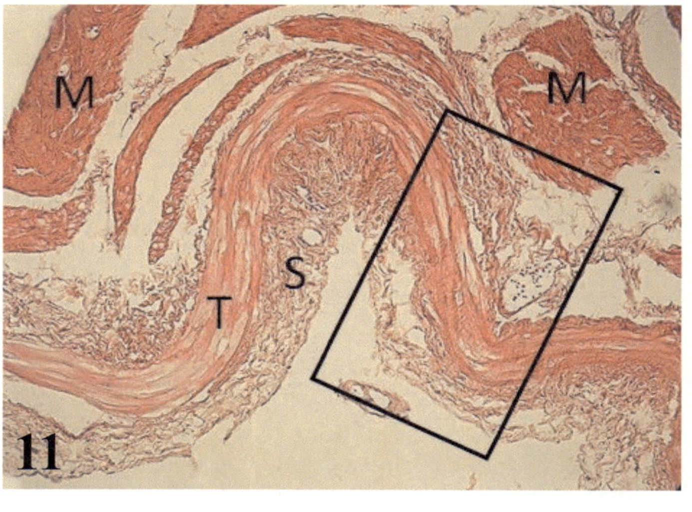

Figure 11.
Histological section of the proventriculus. M: muscular; T: tendon layer and S: serosa. Staining: H-E 200×.

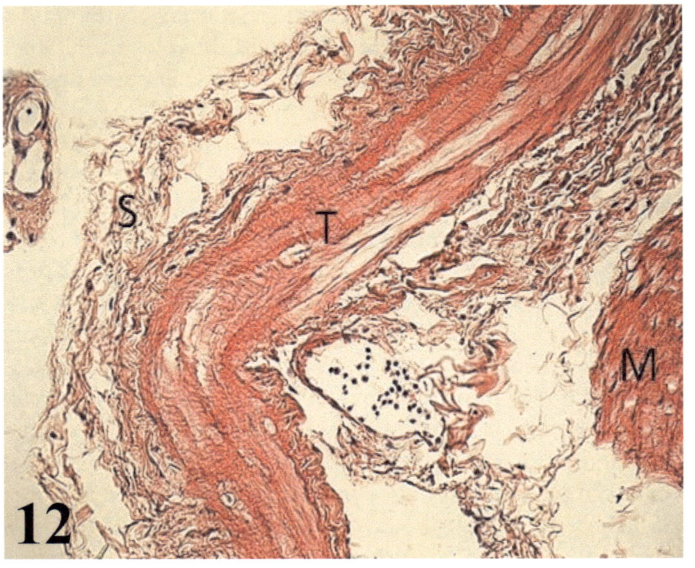

Figure 12.
Higher magnification of the inset of ***Figure 11****. M: muscular; T: tendon layer and S: serosa. Staining: H-E 400×.*

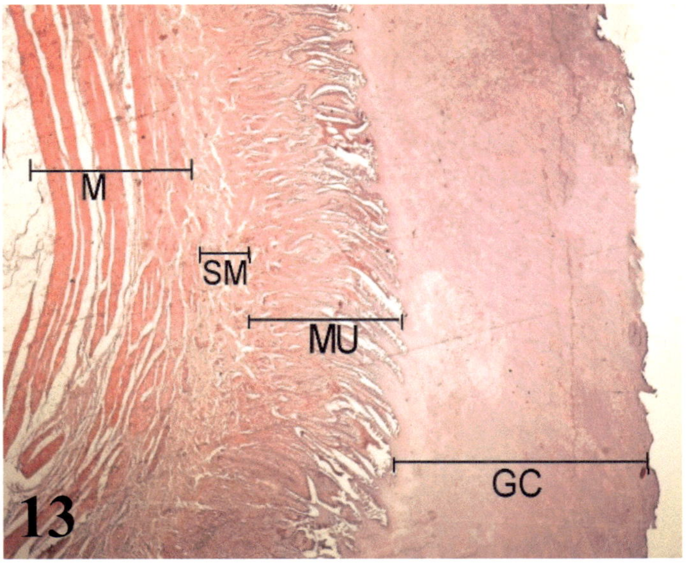

Figure 13.
Panoramic histological section of the muscular stomach wall. GC: glycocalyx; MU: mucosa; SM: submucosa and M: muscular. Staining: H-E 12.5×.

4.1 Small intestine

It consists of three segments called the duodenum, jejunum, and ileum. There are no limits defined between them. Its basic structure is similar to its entire length, being able to find different characteristics according to its function [13].

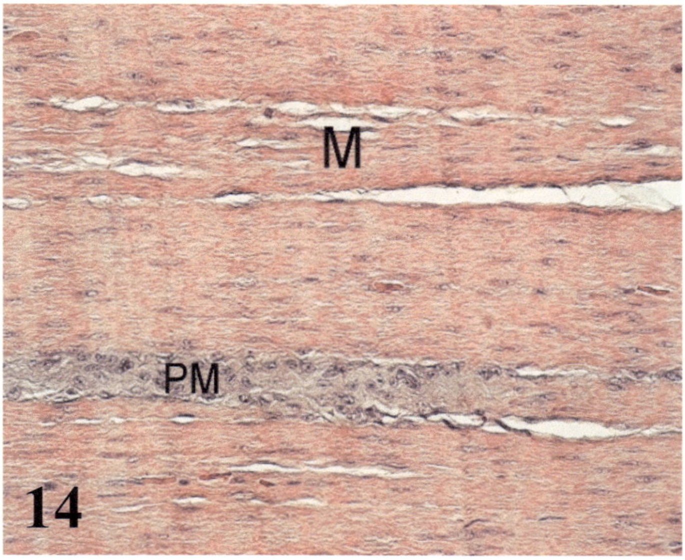

Figure 14.
Histological section of the muscular stomach. M: muscular and PM: myenteric nervous plexus. Staining: H-E 400×.

Duodenum: The mucosa is thick with a simple prismatic epithelium with striated plate and goblet cells forming numerous villi. Numerous intestinal crypts are seen at the base of the villi. Lamina propria of loose connective tissue with lymph nodes. The muscularis mucosa is thin and consists of an internal circular layer and an external longitudinal layer. The inner circular layer branches into the lamina propria to the apex of the villi (**Figures 15** and **16**). The poorly developed submucosa corresponds to loose connective tissue with nerve plexuses. Mucosal and submucosal projections corresponding to connivent valves are observed (**Figures 15** and **16**). The muscularis shows an internal circular layer and a finer external longitudinal layer. Numerous myenteric nerve plexuses are seen between the two layers. The serosa is constituted by the mesothelium and loose connective tissue.

Jejunum: This segment does not present the connivent valves described in the duodenum. However, the basic histological structure of this segment is similar to that of the duodenum. The mucosa with simple prismatic lining epithelium with striated plate and numerous goblet cells show long villi with smooth muscle fibers from the muscularis mucosa (**Figures 17** and **18**). The lamina propria with lymph nodes. The muscularis mucosa is thin and corresponds to two layers of smooth muscle fibers, one internal with a circular arrangement and the other external longitudinal. The submucosa is poorly developed with submucosal nerve plexuses. The muscularis is thicker than seen in the anterior segment, with an inner circular layer and a thinner outer longitudinal layer. The serosa is constituted by the mesothelium and loose connective tissue.

Ileum: The mucosa with a simple prismatic epithelium, with a striated plate and goblet cells, shows fewer, shorter, and thicker villi than in the anterior segments. Numerous intestinal crypts are observed. The intestinal villi are made up of a central axis of loose connective tissue and smooth muscle fibers from branches of the muscularis mucosa (**Figure 19**). The lamina propria shows lymph nodes. The

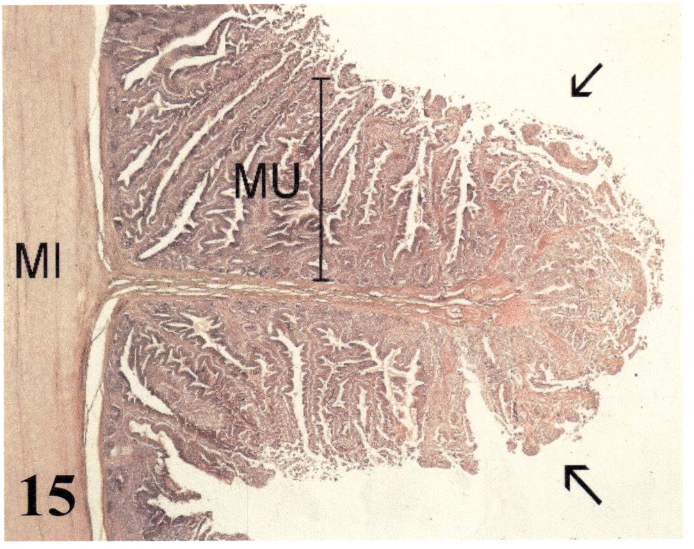

Figure 15.
Histological section of a connivent valve of the duodenum. MU: mucosa; MI: internal muscle and connivent valve (arrows). Staining: H-E 100×.

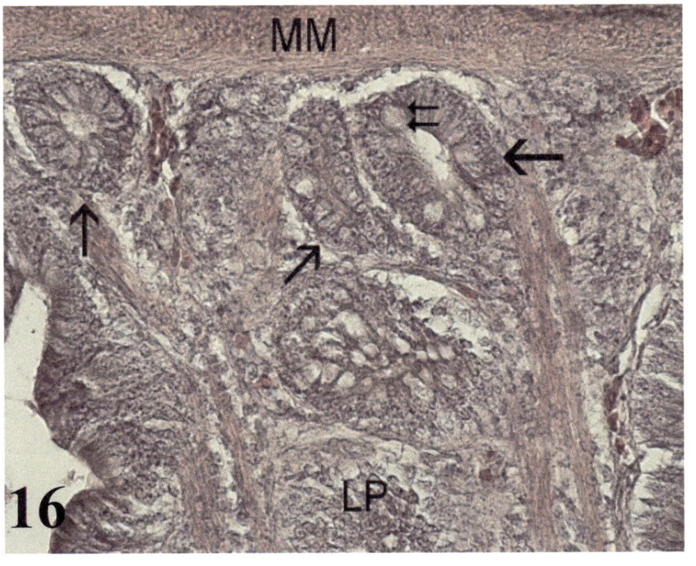

Figure 16.
Histological section of the duodenum. LP: lamina propria; MM: muscular mucosa; intestinal crypts (arrows) and goblet cells (double arrow). Staining: H-E 400×.

muscularis mucosa presents three layers of smooth muscle, an internal circular, an intermediate longitudinal, and an external circular (**Figure 20**). The submucosa is poorly developed with submucosal nerve plexuses. The muscularis is arranged in a thin internal circular layer and a thicker external longitudinal layer, between the two myenteric plexuses are observed. The serosa is made up of the mesothelium and loose connective tissue.

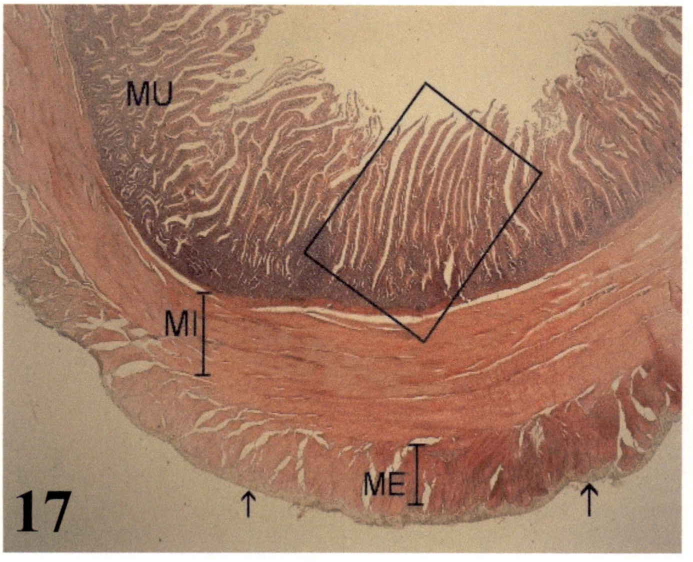

Figure 17.
Histological section of the jejunum. MU: mucosa; MI: internal muscle; ME: external muscle and serosa (arrows). Staining: H-E 40×.

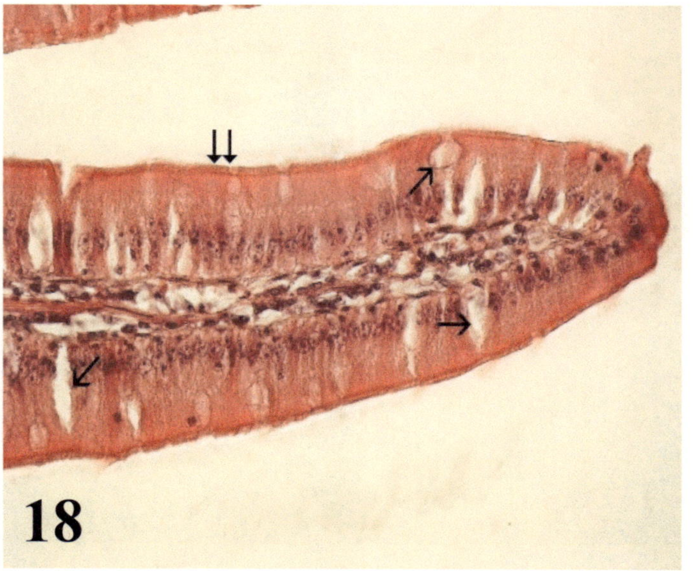

Figure 18.
Higher magnification of the inset of ***Figure 17****. Goblet cells (arrows) and striated plate (double arrow). Staining: H-E 400×.*

4.2 Large intestine

The large intestine of the adult ostrich measures approximately 16 m in length and represents the last portion of the alimentary canal. It has three segments: cecum, colon, and rectum. The general histological structure is similar throughout its length, showing differences in its segments according to its function [13].

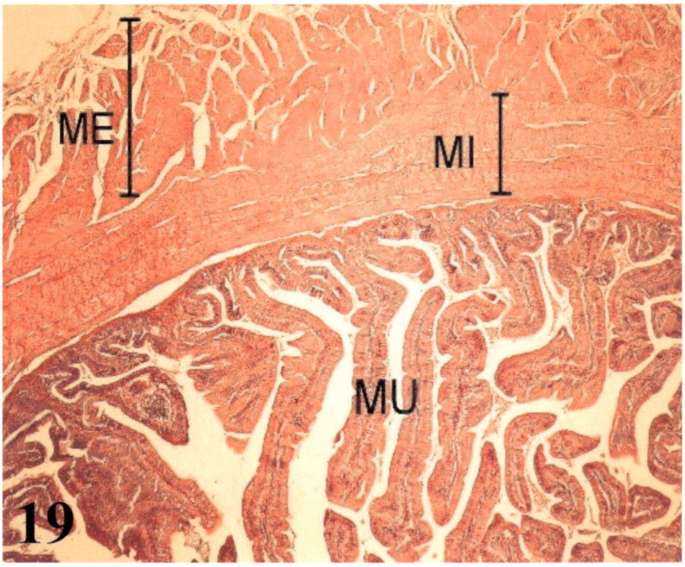

Figure 19.
Histological section of the ileum. MU: mucosa; MI: internal muscle and ME: external muscle. Staining: H-E 100×.

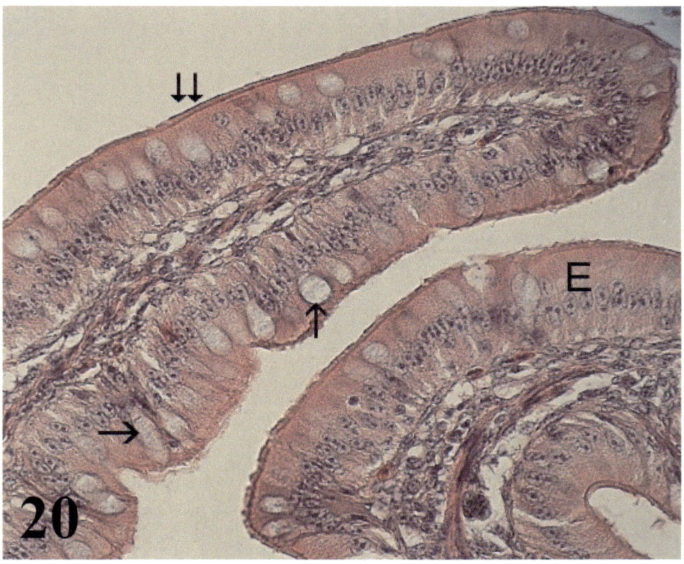

Figure 20.
Histological section of an ileum villus. E: epithelium; Goblet cells (arrows) and striated plate (double arrow). Staining: H-E 400×.

Cecum: The mucosa with a simple prismatic epithelium without intestinal villi. With numerous intestinal crypts and numerous goblet cells, more abundant than in the small intestine. The lamina propria with isolated lymph nodes. The muscularis mucosa corresponds to two layers of smooth muscle fibers, one internal circular and one external longitudinal (**Figure 21**). The submucosa consists of a thin layer of loose connective tissue with submucosal nerve plexuses and a large amount of adipose tissue (**Figure 22**). The muscularis is less developed than in the small intestine, it is

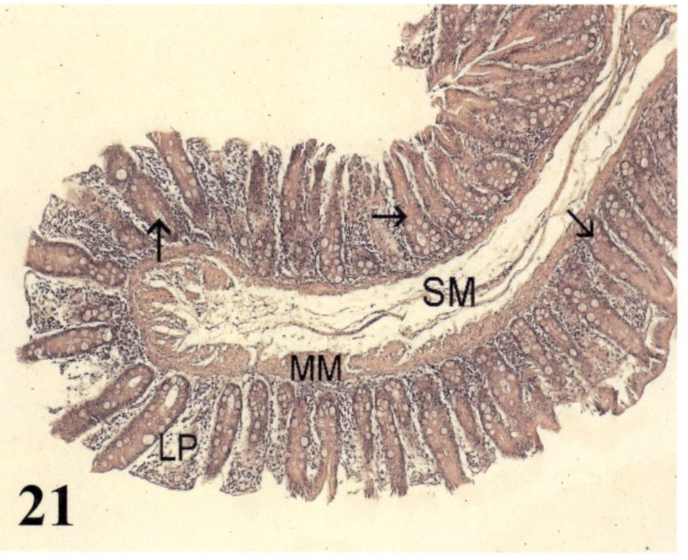

Figure 21.
Histological section of the cecum. LP: lamina propria; MM: muscular mucosa; SM: Submucosa and intestinal crypts (arrows). Staining: H-E 200X.

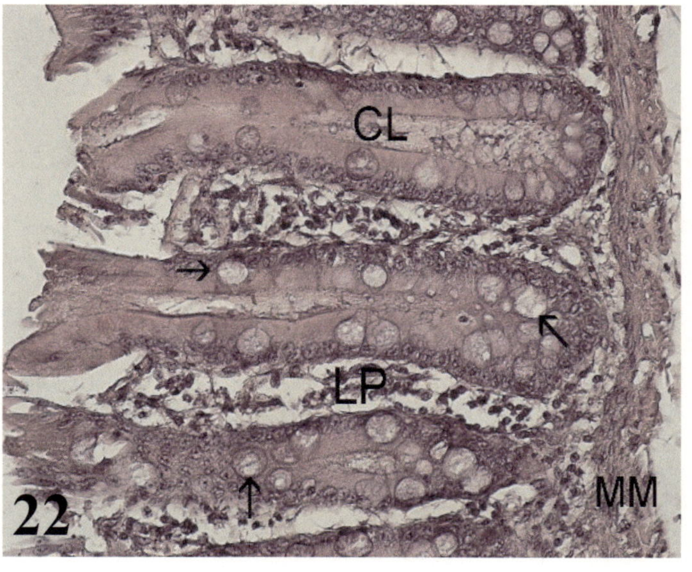

Figure 22.
Histological section of the cecum mucosa. LP: lamina propria; MM: muscular mucosa; CL: intestinal crypts and goblet cells (arrows). Staining: H-E 400×.

formed by a thin internal circular layer and an external longitudinal layer with myenteric nerve plexuses. Adipose tissue is observed between both muscle layers. The serosa is constituted by the mesothelium and loose connective tissue.

Colon: The mucosa with simple prismatic epithelium and fewer intestinal crypts with numerous goblet cells and no intestinal villi. The lamina propria shows isolated lymph nodes (**Figures 23** and **24**). The muscularis mucosa is thin and shows an internal circular layer and an external longitudinal layer. The internal circular layer has ramifications for

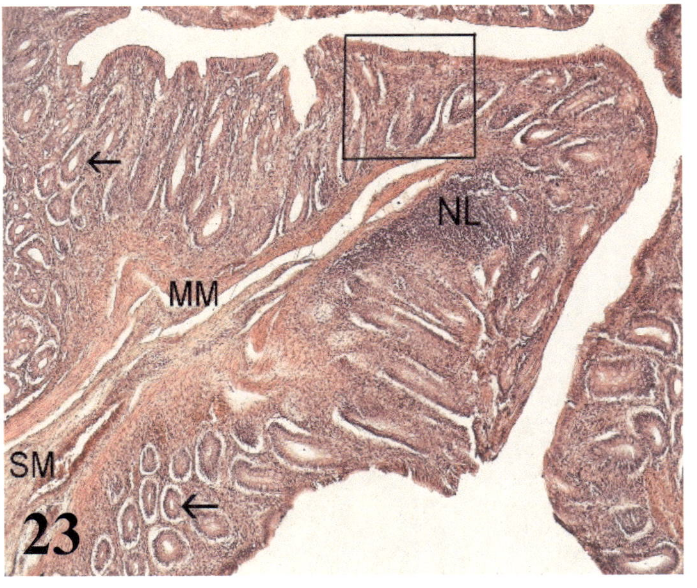

Figure 23.
Histological section of the colon. H-E 200×. MM: muscular mucosa; SM: submucosa; LN: lymph node and intestinal crypts (arrows). Staining: H-E 200×.

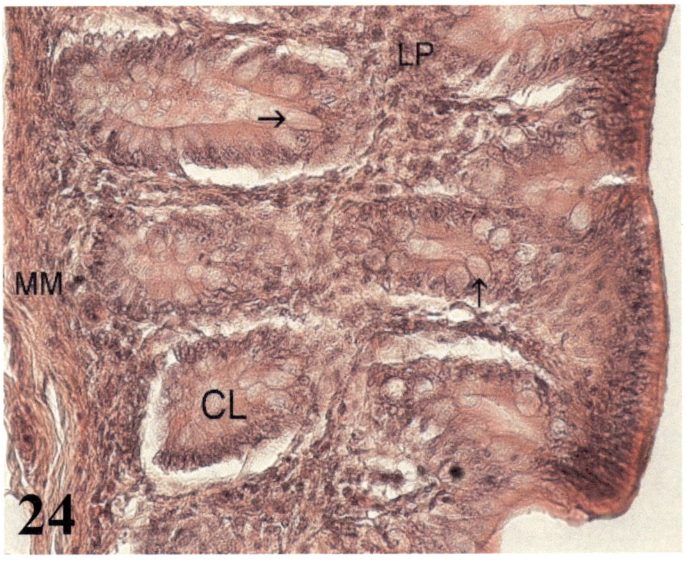

Figure 24.
Higher magnification of the inset of ***Figure 23****. LP: lamina propria; CL: intestinal crypt; MM: muscular mucosa and Goblet cells (arrows). Staining: H-E 400×.*

the lamina propria (**Figure 23**). The submucosa is made up of loose connective tissue with adipose tissue and submucosal nerve plexuses. Connivent valves are observed toward the lumen, which are formed by nuclei of mucosa and submucosa (**Figures 23** and **24**). The muscularis formed by two layers of smooth muscle, an internal one with a circular arrangement and a thinner external one with a longitudinal arrangement with myenteric nerve plexuses. The serous layer of loose connective tissue forms sacs, which follow the direction of the sacs that form in the muscular layer.

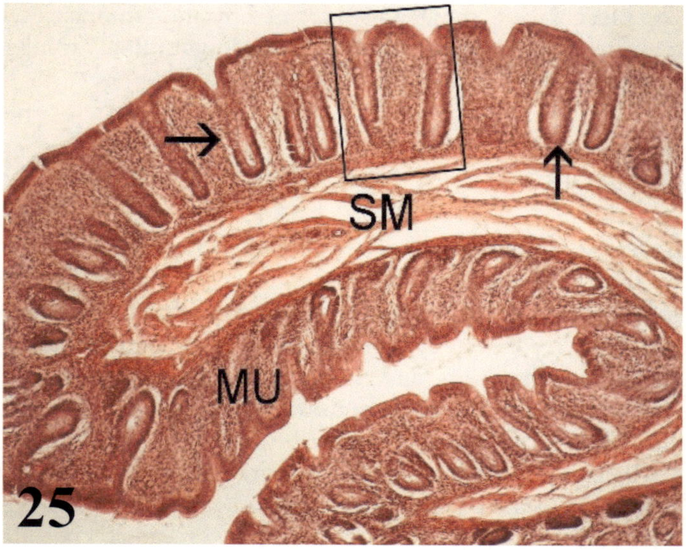

Figure 25.
Histological section of the rectum. MU: mucosa; SM: submucosa and intestinal crypts (arrows). Staining: H-E 200×.

Rectum: Long folds are visible toward the lumen longitudinal in the form of columns, which have a core mucosa and submucosa (**Figure 25**). The mucosa is smooth, devoid of villi, with a simple columnar epithelium. Intestinal crypts are deeper than in the small intestine, with a greater number of cells goblet than in the anterior segments. The muscular of the mucosa is thin and is formed by a layer of smooth muscle fibers in a circular arrangement (**Figure 26**). The submucosa is made up of loose connective tissue. Its structure is similar to that of the other segments

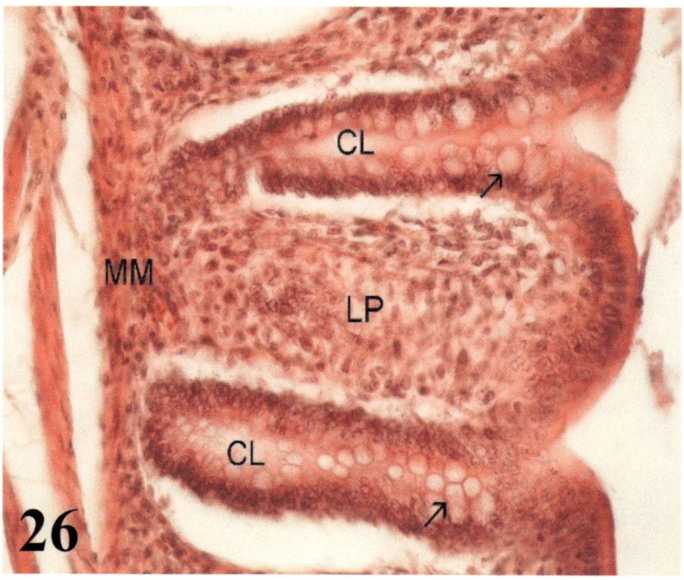

Figure 26.
Higher magnification of the inset of ***Figure*** *25. CL: intestinal crypt; LP: lamina propria; MM: muscular mucosa and goblet cells (arrows). Staining: H-E 400×.*

(**Figure 25**). The muscularis is thick, and is formed by smooth muscles, one internal with a circular arrangement, and the other external longitudinal arrangement. The serosa does not show any differentiation from the usual.

5. Respiratory system

5.1 General description of the respiratory system of the ostrich

The segments analyzed correspond to the larynx, trachea, syrinx, extrapulmonary bronchus, intrapulmonary bronchus or mesobronchus, secondary bronchus, parabronchus or tertiary bronchus, and air capillaries. The wall of the respiratory system presents three concentric layers or tunics: mucosa, submucosa, and adventitia. The greatest difference in the wall of these pathways was observed at the level of the mucosal layer, which expresses the functional characteristics of each segment [14].

Larynx: The mucosa of the larynx, from cranial to caudal, undergoes a change from noncornified squamous stratified epithelium to ciliated prismatic pseudostratified with goblet cells with shallow crypts (**Figures 27** and **28**). The lamina propria of dense connective tissue with simple tubuloalveolar glands of mucus secretion. Between the glands is diffuse dense lymphoid tissue with isolated lymphoid nodules, which are also frequently seen at the onset and end points of epithelial changes (**Figures 27** and **28**). The submucosa is very thin and has loose connective tissue. The adventitia has hyaline cartilage that constitutes the craniolateral portion of the arytenoid cartilage (**Figures 27** and **28**). Toward the caudal region, the laryngeal mucosa is histologically similar and its adventitia shows a hyaline cartilage corresponding to the cricoid cartilage. More externally, it presents a layer of skeletal musculature that surrounds the larynx. Caudally, the larynx is continuous with numerous cartilaginous rings that constitute the beginning of the trachea.

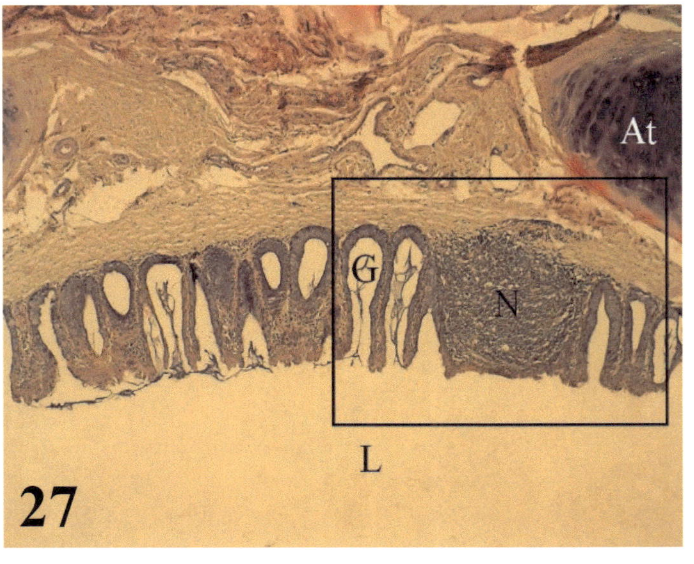

Figure 27.
Ciliated prismatic pseudostratified epithelium with goblet cells, with shallow crypts and simple mucus-secreting tubulo-alveolar glands (G); lymphoid nodule (N) and tracheal cartilage ring (At). Lumen of the larynx (L). Staining: H-E, 100×.

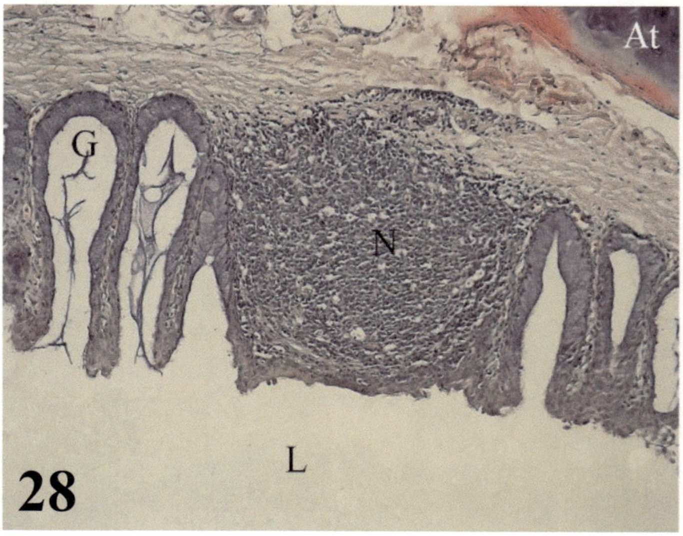

Figure 28.
*Higher magnification of the inset of **Figure** 27. Ciliated prismatic pseudostratified epithelium with goblet cells, with shallow crypts and simple mucus-secreting tubulo-alveolar glands (G), associated with a dense chorion. Lymphoid nodule (N), laryngeal lumen (L), and part of a tracheal ring (At). Staining: H-E, 200×.*

Trachea: The mucosa with ciliated prismatic pseudostratified epithelium with goblet cells with shallow crypts (**Figures 29–31**). The lamina propria of dense connective tissue with simple mucus-secreting tubuloalveolar glands and isolated lymphoid nodules (**Figures 30** and **31**). The loose connective tissue submucosa is sparse. The adventitia is represented by rings of hyaline cartilage. On each side of the trachea and external to the cartilage, fascicles of skeletal muscle can be seen running

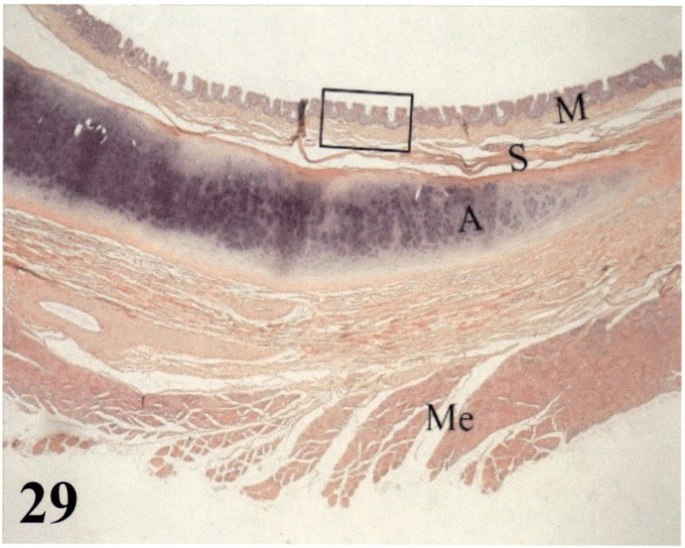

Figure 29.
This figure shows the mucosa (M), submucosa (S), and adventitia A, represented by a tracheal ring. Externally, the skeletal musculature (Me) can be seen. Staining: H-E, 40×.

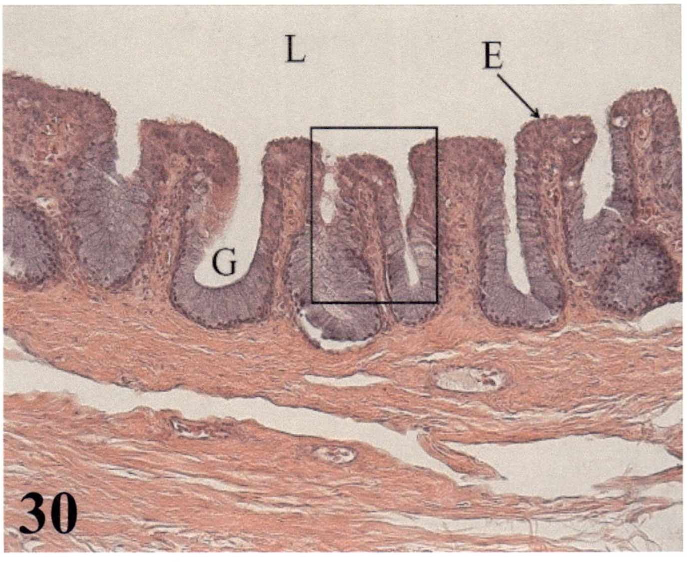

Figure 30.
*Higher magnification of the inset of **Figure 29**. Ciliated prismatic pseudostratified epithelium with goblet cells (E), shallow crypts with simple mucous-secreting tubulo-alveolar glands (G) and chorion of dense connective tissue. Tracheal lumen (L). Staining: H-E, 200×.*

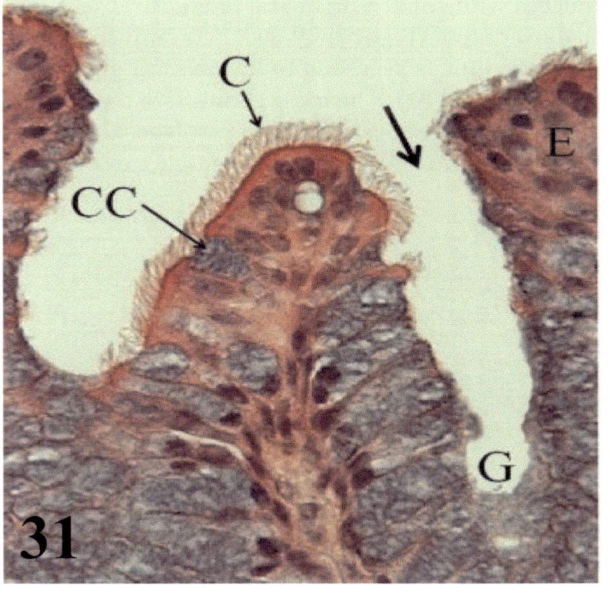

Figure 31.
*Higher magnification of the inset of **Figure 30**. Ciliated prismatic pseudostratified epithelium (E) with goblet cells (CC); shallow crypts (arrow) and simple mucus-secreting tubulo-alveolar glands (G). Cilia (C). Staining: H-E, 1000×.*

longitudinally from the larynx to the syrinx, the lateral muscle of the trachea. The tracheal adventitia at the level of the syrinx is limited on its ventral or anterior face by mesothelium, which constitutes part of the interclavicular air sac.

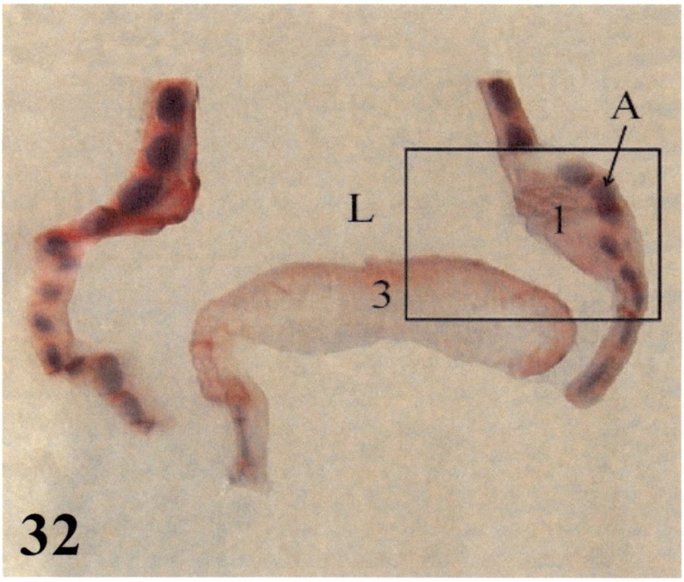

Figure 32.
Coronal panoramic histological section of the syrinx. 1: lateral tympaniform membrane; 3: central membrane or pessulus; A: hyaline cartilage of the tracheosyringeal ring; L: lumen of the syrinx. Staining: H-E, 12.5×.

Syrinx: Microscopically, the syrinx in a coronal section shows the syringeal (tracheosyringeal) rings that are morphologically indistinguishable from the tracheal rings (**Figures 32** and **33**). In the widest area of the syrinx, a group of cartilaginous rings is observed accompanied by abundant loose connective tissue (**Figure 33**). Mucosa with ciliated prismatic pseudostratified epithelium with goblet cells and

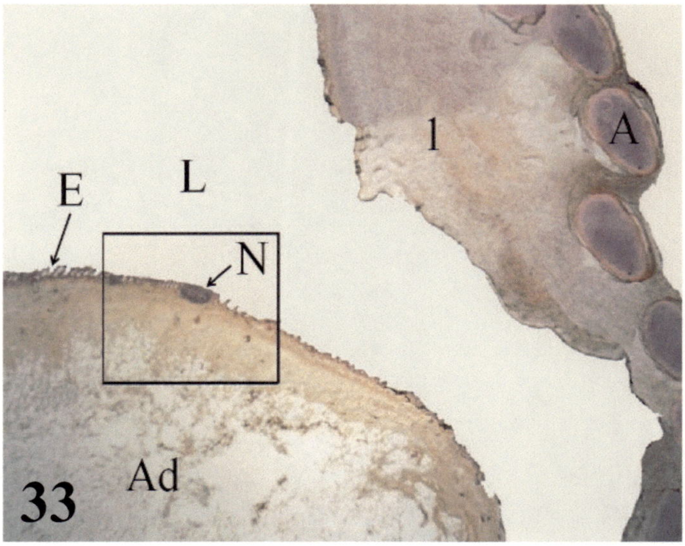

Figure 33.
*Higher magnification of the inset of **Figure 32**. 1: lateral tympani-form membrane; A: Hyaline cartilage of the tracheosyringeal ring. In the pessulus, the mucosa with its underlying epithelium (E) and adipose tissue (Ad) can be seen; lymph node (N). Lumen of the syrinx (L). Staining: H-E, 40×.*

shallow crypts. The lamina propria of dense connective tissue with simple tubuloalveolar glands of mucus secretion, and isolated lymphoid nodules are observed (**Figures 34** and **35**). In the caudal direction, the adventitia is represented by cartilaginous rings that become semirings (bronchosyringeals), which are continuous with those that constitute the extrapulmonary primary bronchi. On the floor of the bifurcation of the syrinx, a double thickening of a medial and dorsoventral arrangement is observed, which prominence toward the syringeal cavity, the pessulus (**Figure 32**), with a ciliated prismatic pseudostratified epithelium with goblet cells and shallow crypts and lamina propria of connective tissue dense with simple tubuloalveolar glands of mucus secretion with isolated lymphoid nodules and deeper a thick layer of adipose tissue (**Figures 33–35**). The loose connective tissue submucosa is sparse. The hyaline cartilage adventitia in the syrinx is limited by the pleural mesothelium, which becomes the wall of the thoracoabdominal air sacs, the latter present a ciliated prismatic pseudostratified epithelium with goblet cells, which in some sectors can be seen simple tubuloalveolar glands of mucous secretion (**Figures 36–38**).

Extrapulmonary primary bronchi: They originate from the bifurcation of the syrinx. They present an extrapulmonary portion and another intrapulmonary or mesobronchus, they have a semi-ring of hyaline cartilage, whose ends are joined in the medial region by connective tissue and smooth muscle. Microscopically, the extrapulmonary primary bronchi maintain the histological characteristics of the tracheal wall (**Figures 36** and **37**). Mucosa with ciliated prismatic pseudostratified epithelium with goblet cells. Lamina propria of dense connective tissue with isolated lymphoid nodules. The thin submucosa of loose connective tissue. The adventitia is formed by semi-rings of hyaline cartilage, whose ends are joined in their medial region by smooth muscle and connective tissue. In this region, simple tubuloalveolar glands of mucous secretion are observed. Medially and at the bifurcation of the right and left bronchi, the mucosa constitutes the medial tympani-form membranes (**Figures 32**

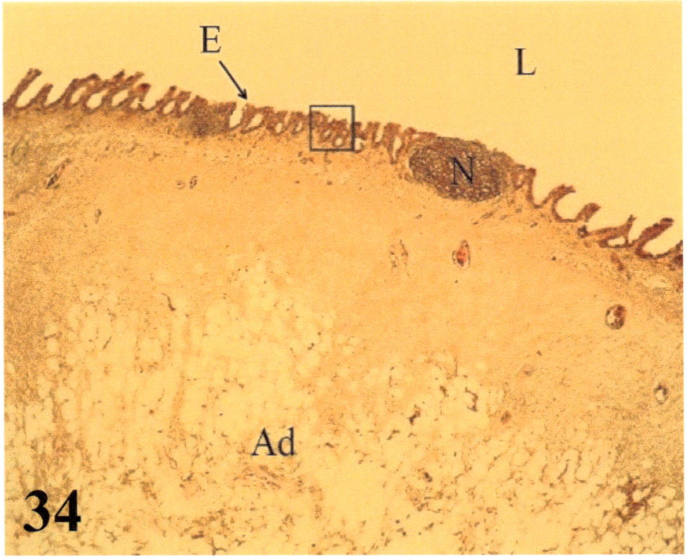

Figure 34.
*Higher magnification of the inset of **Figure 33**. E: ciliated pseudostratified prismatic epithelium with goblet cells, shallow crypts with simple mucus-secreting tubulo-alveolar glands; Ad: adipose tissue; N: lymph node. L: Lumen of the syrinx. Staining: H-E, 100×.*

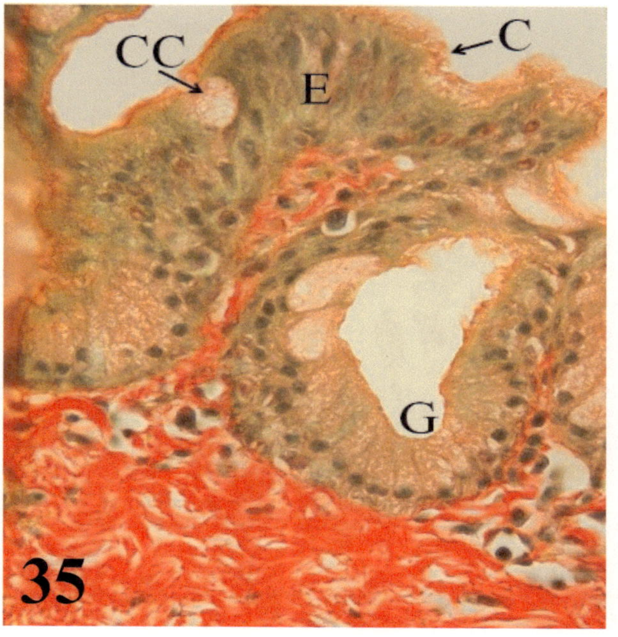

Figure 35.
*Higher magnification of the inset of **Figure 34**. E: Ciliated prismatic pseudostratified epithelium with goblet cells (CC), shallow crypts, and simple mucus-secreting tubulo-alveolar glands (G), with a chorion of dense connective tissue. C: cilia. Staining: Van Gieson, 1000×.*

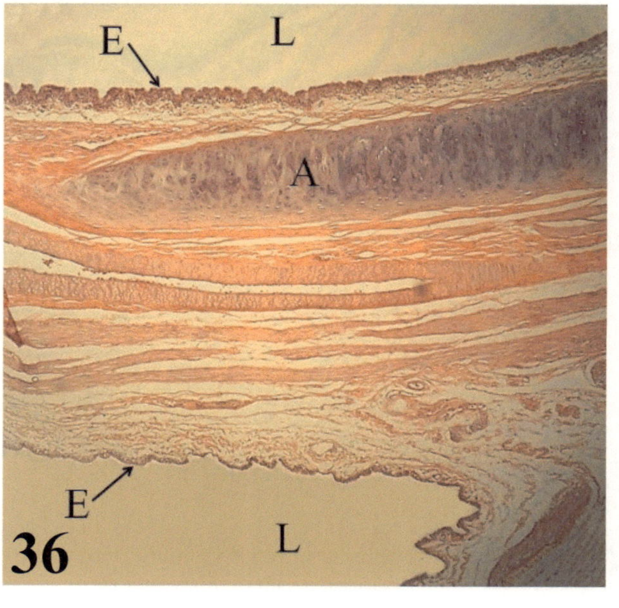

Figure 36.
Extrapulmonary primary bronchus. E: ciliated prismatic pseudostratified epithelium with goblet cells, both in the bronchial (upper) and saccular (lower) lumen (L); A: semi-ring of hyaline cartilage. Staining: H-E, 200×.

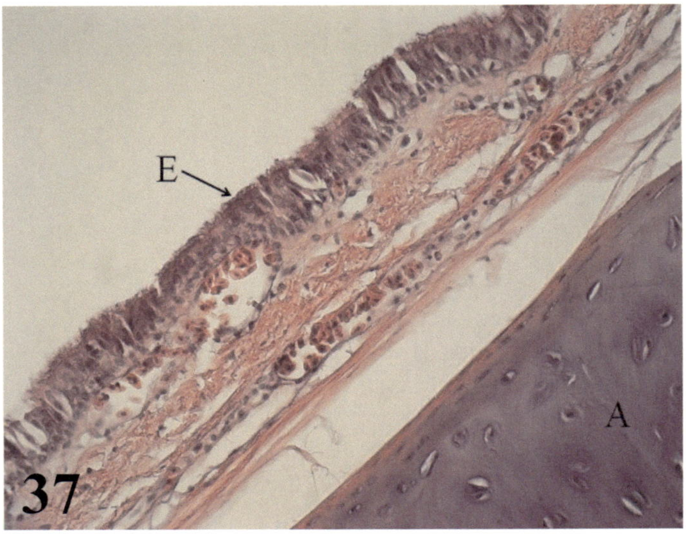

Figure 37.
Extrapulmonary primary bronchus. E: ciliated prismatic pseudostratified epithelium with goblet cells and A: hyaline cartilage of the bronchial semiring. Staining: H-E, 400×.

and **33**). The mucosa and submucosa at this level present numerous folds of different sizes made up of abundant connective tissue interspersed with smooth muscle tissue. Bundles of circularly arranged smooth muscle unite their ends, throughout their entire length. The extrapulmonary primary bronchi are limited lateroventrally by the pleural mesothelium that constitutes the thoracoabdominal air sacs (**Figures 36** and **38**). The portion seen from part of a thoracoabdominal air sac shows a ciliated prismatic pseudostratified lining epithelium with goblet cells (**Figures 36** and **37**).

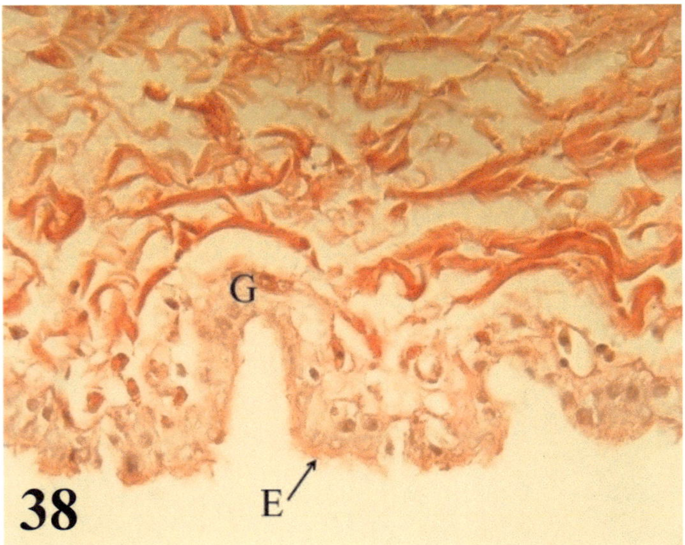

Figure 38.
Wall of the air sac. E: pseudostratified prismatic ciliated epithelium with goblet cells with simple tubuloalveolar glands that secrete mucus (G). Type I collagen tendon layer. Staining: Van Gieson, 1000×.

Air sacks: The air sacs that are in contact with the extrapulmonary primary bronchi present a ciliated prismatic pseudostratified epithelium with cell goblet. However, in some sectors, it can be observed simple secretory tubuloalveolar glands have mucous membranes. Then a thin layer of loose connective tissue, followed by a thick tendon layer of collagen fibers type I (**Figure 38**).

5.2 Lung

In the lungs, we find primary intrapulmonary bronchi or mesobronchi, secondary bronchi distributed into three groups (dorsal, ventral, and lateral), parabronchi or tertiary bronchi, and the lung parenchyma made up of blood capillaries that surround the air capillaries that originate from the atria. This parenchyma is the place where hematosis occurs [14].

Primary intrapulmonary bronchi or mesobronchi: They run medially and craniocaudally throughout the lung until they contact the posterior air sacs through the lateral secondary bronchi. Microscopically: The mucosa with ciliated prismatic pseudostratified epithelium with goblet cells, presents numerous folds with shallow crypts. Lamina propria of dense connective tissue with simple tubuloalveolar glands of mucous secretion throughout its entire length. The submucosa of loose connective tissue with abundant bundles of smooth muscle fibers. The adventitia with hyaline cartilage is arranged in cartilaginous plates (not in semi-rings), a characteristic that gives its mucosa a scalloped appearance. These plates decrease in size until they disappear at the level of the secondary bronchi. This adventitia is surrounded by a layer of smooth muscle with a helical arrangement and its ventral region is surrounded by the pleural serosa (**Figures 39** and **40**).

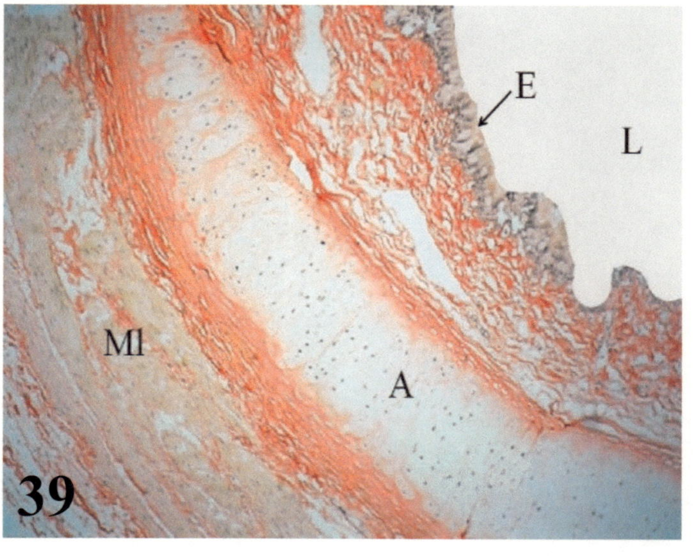

Figure 39.
Primary intrapulmonary bronchus or mesobronchus. E: Ciliated prismatic pseudostratified epithelium with goblet cells, with simple tubulo-alveolar glands and chorion of dense connective tissue; A: hyaline cartilage; ML: smooth muscle. Staining: Van Gieson, 200×.

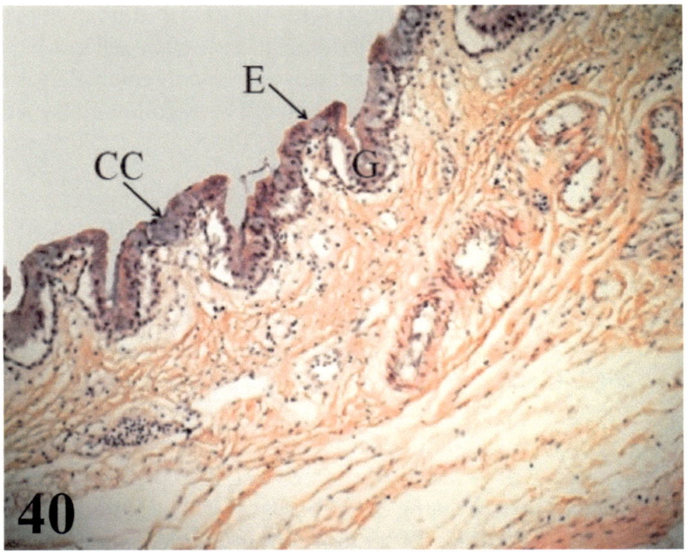

Figure 40.
Primary intrapulmonary bronchus or mesobronchus. E: ciliated prismatic pseudostratified epithelium with goblet cells (CC), deep crypts with simple tubulo-alveolar glands of mucus secretion (G). Staining: H-E, 200×.

Secondary bronchi: In each lung, the intrapulmonary primary bronchi or mesobronchi give rise to three groups of secondary bronchi: four dorsal secondaries, three ventral secondaries, and two lateral secondaries. These branch in the lung parenchyma and give rise to the parabronchi or tertiary bronchi (**Figures 41** and **42**). The mucosa of the secondary bronchi with a ciliated prismatic pseudostratified epithelium with goblet cells and shallow crypts. Lamina propria of dense connective

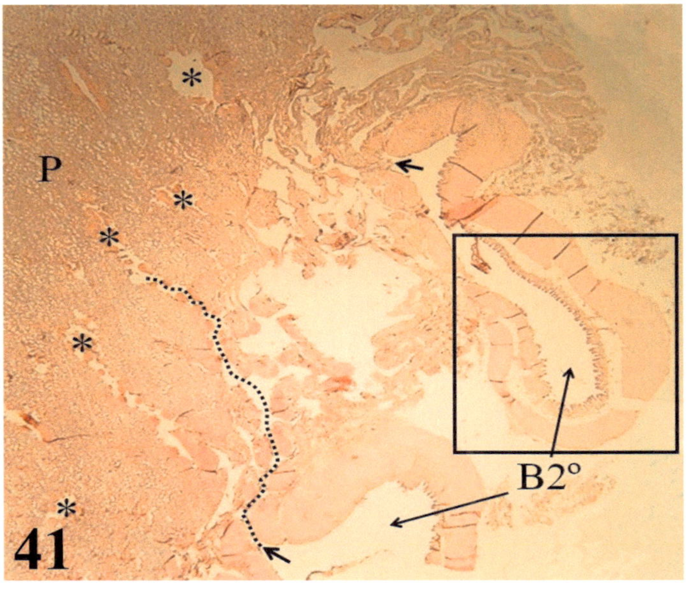

Figure 41.
Panoramic histological section showing the lung parenchyma (P), parabronchi (*); and its origin (arrows) from the secondary bronchi (B2°). The dashed line marks the luminal path. Stain: H-E, 12.5×.

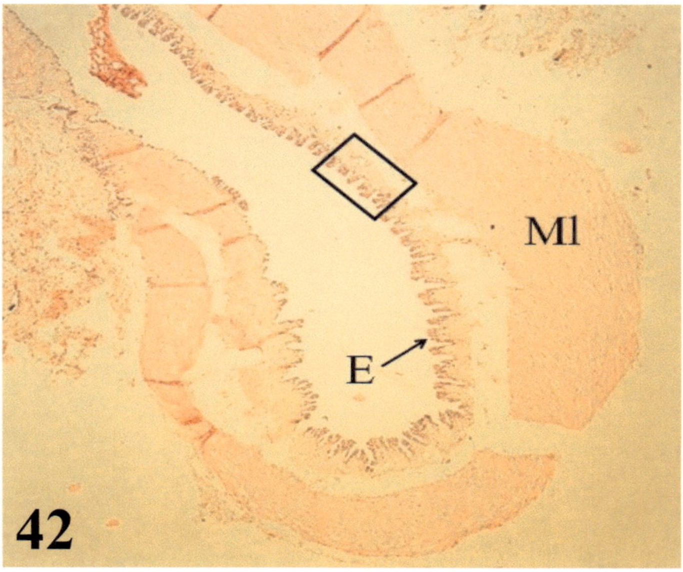

Figure 42.
*Higher magnification of the inset of **Figure 41**. Histological section of the secondary bronchus. E: ciliated prismatic pseudostratified epithelium with goblet cells, shallow crypts, and simple mucus-secreting tubuloalveolar glands and smooth muscle (ML). Staining: HE, 100×.*

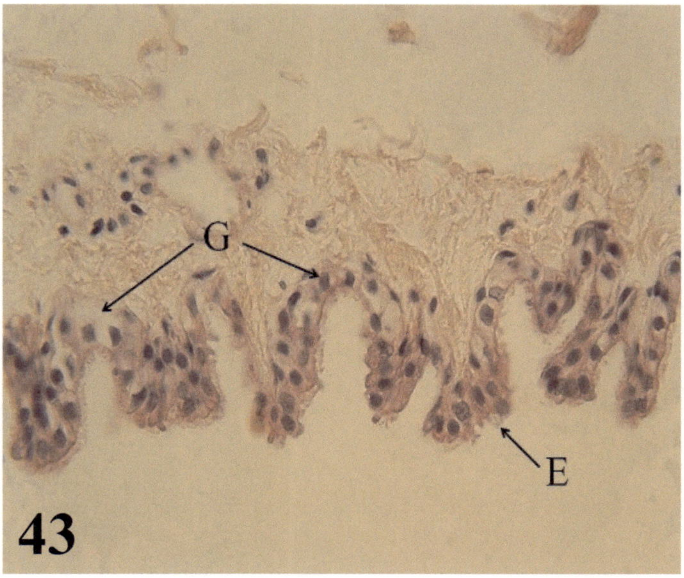

Figure 43.
*Higher magnification of the inset of **Figure 42**. E: pseudostratified prismatic ciliated epithelium with goblet cells, shallow crypts, and simple mucus-secreting tubulo-alveolar glands (G). Staining: H-E, 1000×.*

tissue with simple mucus-secreting tubuloalveolar glands. Under the lamina propria, there is a thick layer of smooth muscle in a circular arrangement. In the adventitia of the secondary bronchi, small plates of hyaline cartilage are still observed that disappear before originating the parabronchi (**Figures 41, 43,** and **44**).

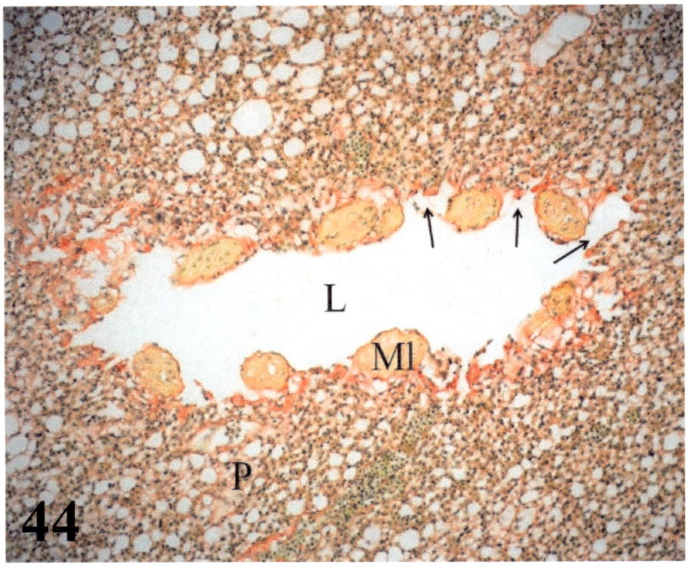

Figure 44.
Histological section of the lung parenchyma (P) and of a transversely arranged parabronchi or tertiary bronchus with its lumen (L); smooth muscle (Ml) and atria (arrows). Staining: Van Gieson, 200×.

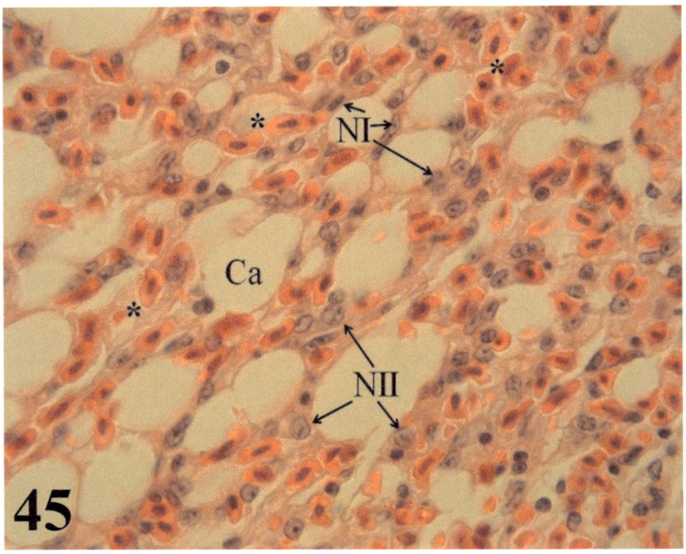

Figure 45.
Histological section of the lung parenchyma. Aerial capillaries (Ca) with flat cells (NI) and cuboidal cells (NII), surrounded by blood capillaries with nucleated erythrocytes (*). Staining: H-E, 1000×.

Parabronchi or tertiary bronchi: The mucosa with a low cubic simple epithelium, lamina propria formed by a thin layer of connective tissue. The parabronchi are characterized by having bundles of smooth muscle with a helical arrangement, both clockwise and counterclockwise. Multiple cavities called atria originate from the parabronchi, which are continuous with the air capillaries, both lined by a simple flat epithelium (**Figures 41** and **42**). Surrounding the wall of the atrium and around the bundles of smooth muscle, there are accumulations of macrophages, among these structures are numerous lymphocytes, blood, and lymphatic capillaries.

Air capillaries: These originate and branch from the atria, which in turn, arise from the parabronchi. Microscopically, together with the blood capillaries, they constitute the lung parenchyma (**Figure 45**). With simple epithelium, which presents flat cells (type I pneumocytes) and others cubic (type II pneumocytes). These air capillaries are closely related to numerous blood capillaries, linked by a common basement membrane. In the wall of the atria, which originate from the parabronchi and constitute the entrance to the air capillaries, accumulations of fixed and free-form macrophages are observed in their lumen (**Figure 45**).

6. Female reproductive system

The female ostrich presents only the left ovary and oviduct. The ovary is large, cluster-shaped, which varies according to the seasons. It presents primordial, previtellogenic, vitellogenic, and atretic follicles. The oviduct has the following segments from the cephalad to the caudal: infundibulum, magnum, isthmus, uterus, and vagina, which empties into the cloaca. In them, there are folds of varying length, thickness, and number that compromise the mucosa and submucosa, with glands of mucous and serous secretion with the exception of the vagina, [15].

6.1 Ovary: ovarian follicles in different stages of maturation are arranged as a cluster

The ovarian cortex is covered in some areas by simple cuboidal epithelium, in others by ciliated simple prismatic epithelium and with pseudostratified prismatic epithelium (**Figure 46**). Beneath the epithelium is a thin layer of dense connective tissue, the albuginea (**Figure 46**). In the cortical stroma, follicles are observed in different stages of maturation, some of them, the largest, protrude on the surface. These are classified as primordial, previtellogenic, vitellogenic, and atresic. In primordial follicles, the oocyte with a vesicular nucleus is surrounded by a flat or low cuboidal epithelium of follicular cells that correspond to the granulosa layer. Externally to the basement membrane, some myofibroblasts are arranged (**Figure 47**). In previtellogenic follicles, the oocyte with a vesicular nucleus and the scant yolk is surrounded by a granulosa of cuboidal epithelium, a very conspicuous basement membrane and externally an undifferentiated theca vascularized with concentrically arranged myofibroblasts (**Figures 46** and **47**). In vitellogenic follicles, the oocyte with abundant yolk is surrounded by a granular layer of cuboidal epithelium, the theca interna vascularized, and the theca externa with abundant myofibroblasts arranged concentrically (**Figures 48** and **49**). Atretic follicles, which are smaller, have numerous lipid vacuoles.

6.2 Oviduct: it is a duct that has five segments

Infundibulum: it is the first segment of the oviduct, adjacent to the thin-walled ovary with long and numerous folds. The mucosa with ciliated pseudostratified prismatic epithelium with goblet cells and a lamina propria of dense connective tissue with abundant collagenous fibers. Interposed in the epithelium are simple tubuloalveolar (**Figures 50** and **51**) and simple alveolar glands of mucous secretion.

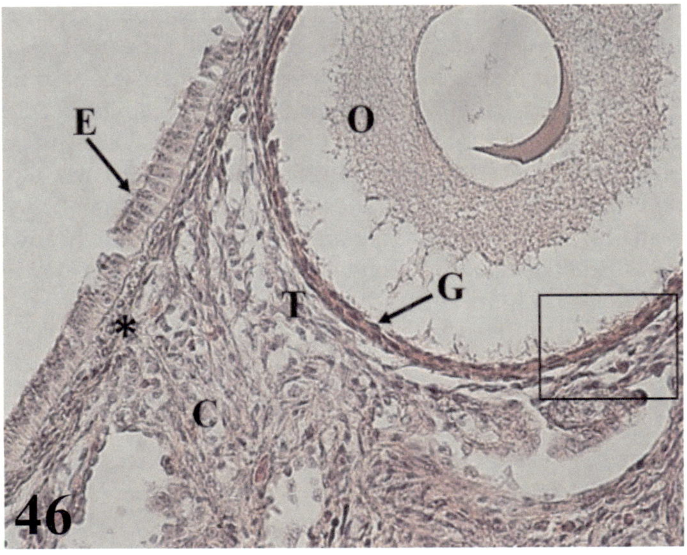

Figure 46.
Histological section of the ovary. E: simple ciliated prismatic ovarian epithelium; O: oocyte from a previtellogenic follicle; C: chorion of connective tissue; (*): albugínea; T: highly vascularized undifferentiated theca and G: granulosa. Staining: H-E, 400×.

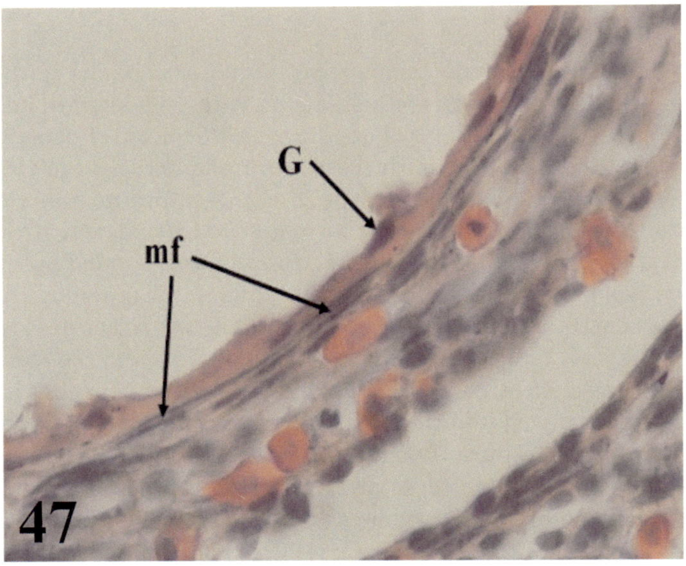

Figure 47.
Higher magnification of the inset of ***Figure 46****. Highly vascularized undifferentiated theca with concentrically arranged myofibroblasts (mf) surrounding the granulosa basement membrane (G) of a previtellogenic follicle. Staining: H-E, 1000×.*

No submucosa is seen. The muscularis, plexiform in arrangement, emits smooth muscle fibers that project through the mucosa of the folds. The serosa possesses dense connective tissue and is surrounded by mesothelium.

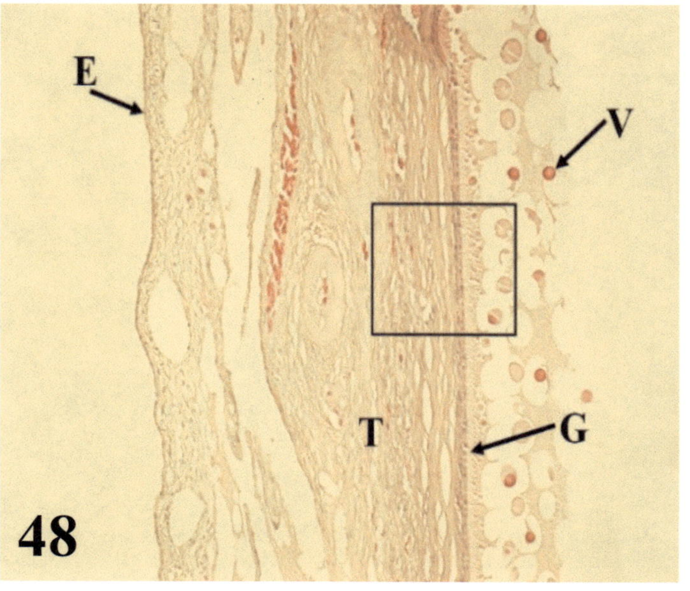

Figure 48.
Panoramic histological section of the wall of a vitellogenic follicle. E: ovarian epithelium; T: theca; G: granulosa and V: drops of yolk. Staining: H-E, 200×.

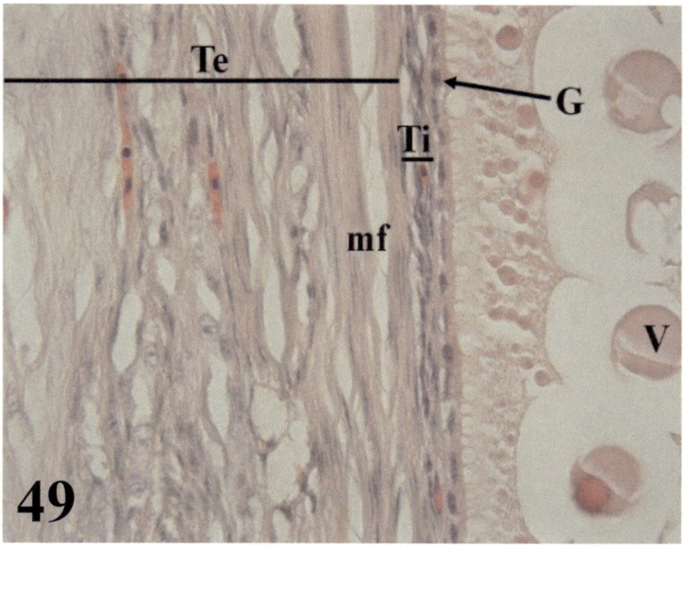

Figure 49.
Higher magnification of the inset of ***Figure 48****. Te: theca externa; mf: with myofibroblasts; Ti: internal theca; G: granulosa and V: drops of yolk. Staining: H-E, 1000×.*

Magnum: it is the longest segment of the oviduct, with the greatest diameter and numerous thick folds. Mucosa with ciliated prismatic pseudostratified epithelium with goblet cells. The lamina propria with abundant branched tubular glands of serous secretion (**Figures 52** and **53**). The loose connective tissue submucosa is highly irrigated and with smooth muscle fibers. The muscularis is arranged in an internal

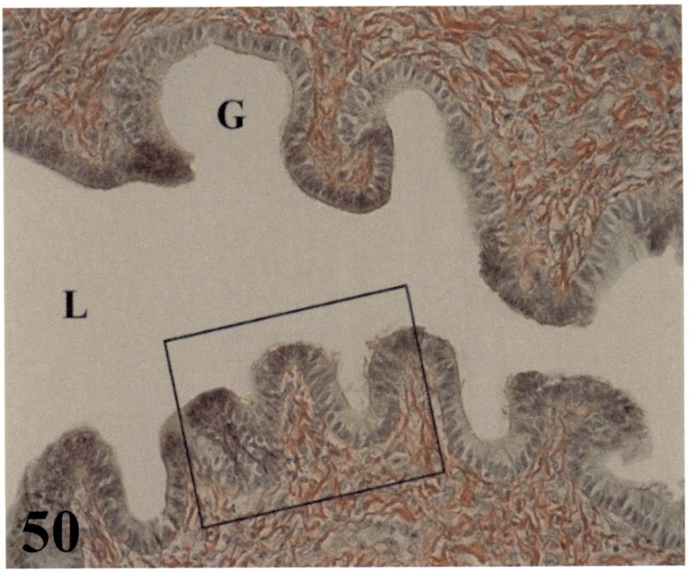

Figure 50.
Shows the mucosa of the infundibulum. L: lumen and G: simple alveolar glands of mucus secretion. Staining: Van Gieson, 400×.

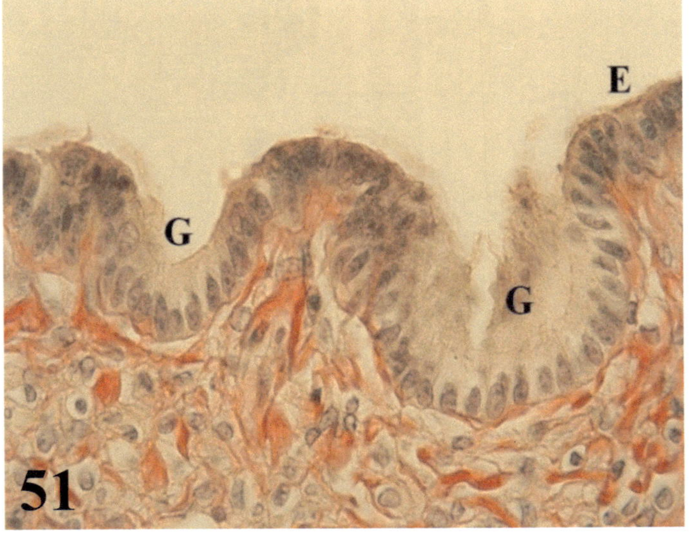

Figure 51.
*Higher magnification of the inset of **Figure 50**. G: simple alveolar glands and E: ciliated pseudostratified prismatic epithelium with mucosal goblet cells. Staining: Van Gieson, 1000×.*

circular layer that is projected slightly by the wide submucosa of the folds and an external layer of plexiform arrangement. The serosa has loose connective tissue that is highly irrigated and is surrounded by mesothelium.

Isthmus: it is the shortest segment of the oviduct, with the smallest diameter and thinnest wall, with numerous tortuous folds. Mucosa with ciliated pseudostratified prismatic epithelium with goblet cells. The lamina propria with abundant simple tubular glands of serous secretion (**Figures 54** and **55**). The submucosa, made of loose

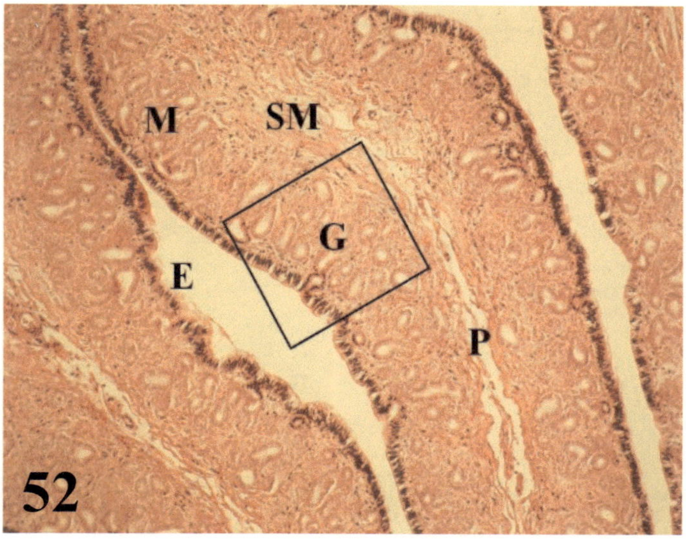

Figure 52.
The figure shows a fold of the magnum (P), the mucosa (M) with its epithelium (E), chorion of connective tissue with glands (G) and submucosa (SM). Staining: HE, 200×.

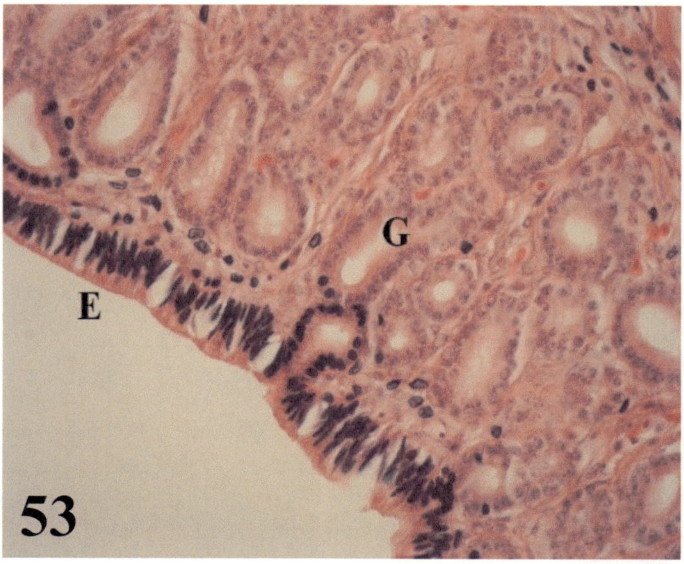

Figure 53.
*The figure shows the inset of **Figure 52** at higher magnification. E: ciliated pseudostratified prismatic lining epithelium with goblet cells and G: chorion of connective tissue with branched tubular glands of serous secretion. Staining: H-E, 400×.*

connective tissue, is well supplied with some smooth muscle fibers. The muscularis is arranged in an internal circular layer that is projected finely by the submucosa of the folds and an external layer of plexiform arrangement. The serosa has loose connective tissue with a high blood supply and is surrounded by mesothelium.

Uterus: globose in shape with numerous long folds (**Figure 56**). Mucosa with ciliated pseudostratified prismatic epithelium with goblet cells. The lamina propria shows abundant branched tubular glands of serous secretion (**Figures 56** and **57**). The

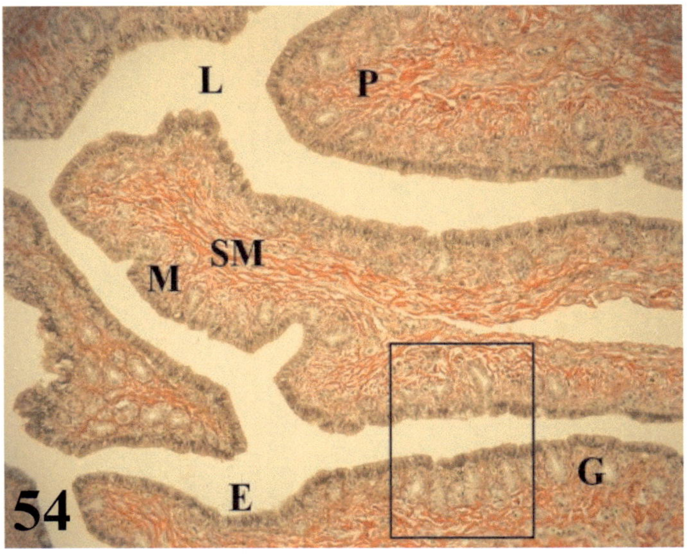

Figure 54.
The figure shows a fold of the isthmus (P), the mucosa (M) with its epithelium (E), chorion of dense connective tissue with glands (G), submucosa (SM) and lumen (L). Staining: Van Gieson, 200×.

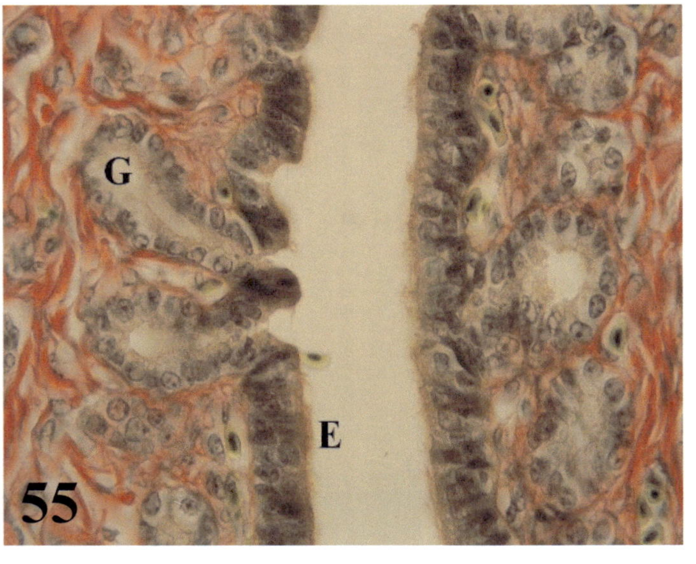

Figure 55.
Higher magnification of the inset of **Figure 54***. E: ciliated pseudostratified prismatic lining epithelium with goblet cells and G: chorion of dense connective tissue with simple tubular glands of serous secretion. Staining: Van Gieson, 1000×X.*

thin submucosa of loose connective tissue is well supplied with some smooth muscle fibers. The muscularis is arranged in an internal circular layer that emits fine projections toward the submucosa and an external layer with a plexiform arrangement. The serosa contains loose connective tissue with a high blood supply and is surrounded by mesothelium.

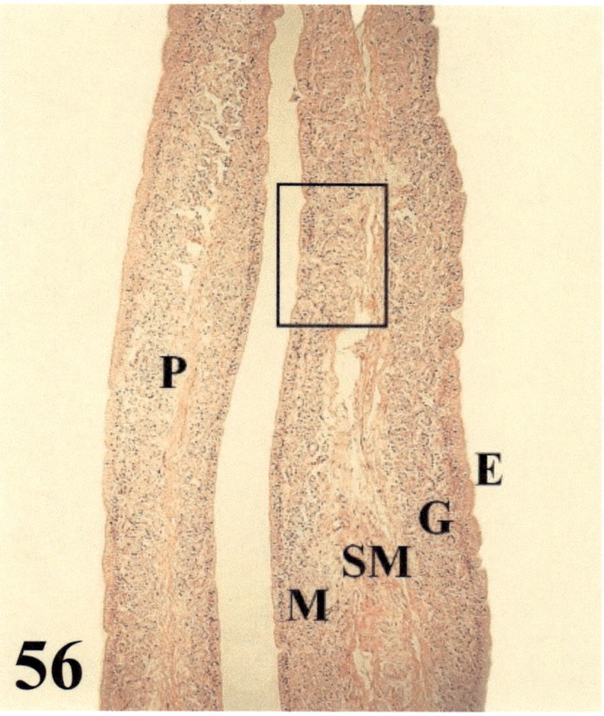

Figure 56.
The figure shows a fold of the uterus (P), the mucosa (M) with its epithelium (E), and chorion of connective tissue with numerous glands (G) and submucosa (SM). Staining: H-E, 200×.

Vagina: it is a short and narrow segment with a highly developed wall with numerous, long, and highly branched folds, with numerous villi forming a dense pattern (**Figures 58** and **59**). Mucosa presents ciliated pseudostratified prismatic epithelium with goblet cells (**Figure 60**). The dense connective tissue lamina propria is glandless. The arrangement of these folds gives rise to the formation of numerous spaces. The loose connective tissue submucosa is highly irrigated and bundled with smooth muscle fibers. The muscularis is arranged in a thick internal circular layer that projects notably through the submucosa of the folds and its ramifications, also reaching the villi and an external layer with a plexiform arrangement (**Figures 58** and **59**). An adventitia of loose connective tissue and a serosa that has loose connective tissue and is surrounded by mesothelium.

7. Discussion and conclusions

At the level of the glands attached to the ostrich digestive system, certain histological differences are observed. The ostrich liver does not present significant differences from the liver of the chicken and other species [16]. However, the arrangement of the hepatocyte cords, both around the central vein and the portal spaces, is radiated, and a clear organization into lobules is not observed. In the ostrich proventriculus, we can point out that it does not present great differences from that of the chicken [16, 17]. Although the simple and branched tubular glands of the mucosa are poorly developed, they are very similar to the fundic glands of monogastric mammals (dogs)

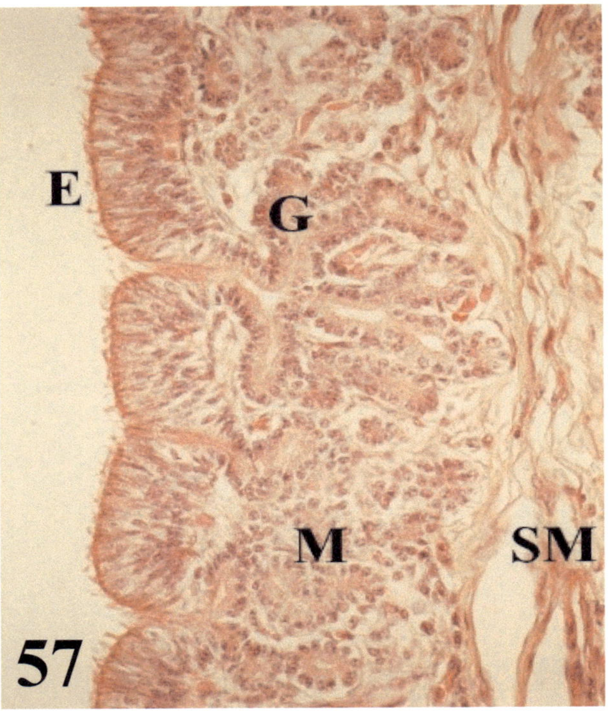

Figure 57.
*Higher magnification of the inset of **Figure 56**. Mucosa (M) with ciliated pseudostratified prismatic lining epithelium with goblet cells (E), connective tissue chorion with branching tubular glands of serous secretion (G), and submucosa (SM). Staining: H-E, 400×.*

and non-true ruminants (camelids) [18, 19]. The ostrich and chicken submucosa are similar, presenting compound tubulo-alveolar glands [16, 17]. In the ostrich, it should be noted that these glands have a large number of parietal cells that secrete hydrochloric acid [12]. The pancreas of both the ostrich and the chicken present adenomeres of the tubulo-acinar type composed of serous secretion [16].

The epithelium of the esophageal mucosa in the ostrich is non-keratinized stratified squamous type, while in ruminants the mucosa presents a high degree of parakeratinization, which is slight in the horse [20]. The lamina propria in the ostrich shows numerous simple and branched tubuloalveolar glands of mucus secretion, arranged along its entire length [13]. The hen presents mucous secretion glands of the tubuloacinar type composed [21]. In other species, such as ruminants, horses and pigs, the mucosa lacks these glands [16]. The muscularis mucosa in the ostrich is observed as a thick layer of longitudinally arranged smooth muscle fibers. The hen has two layers, an internal one with a circular arrangement, formed by thin fascicles that are not always appreciable, and an external longitudinal one that is more developed [22]. The muscularis mucosa of domestic birds is formed only by a layer of longitudinally arranged smooth muscle fibers [21]. In the ostrich, the esophageal submucosa is very poorly developed, and as in the chicken, it lacks glands [21], while in the horse and ruminants, these glands are present, but they are only observed in the pharyngoesophageal junction [16, 20]. The muscularis of the ostrich esophagus, as in the chicken, has two layers of smooth muscle with an internal circular and external longitudinal arrangement [21, 22]. In the ruminant it is made up of striated

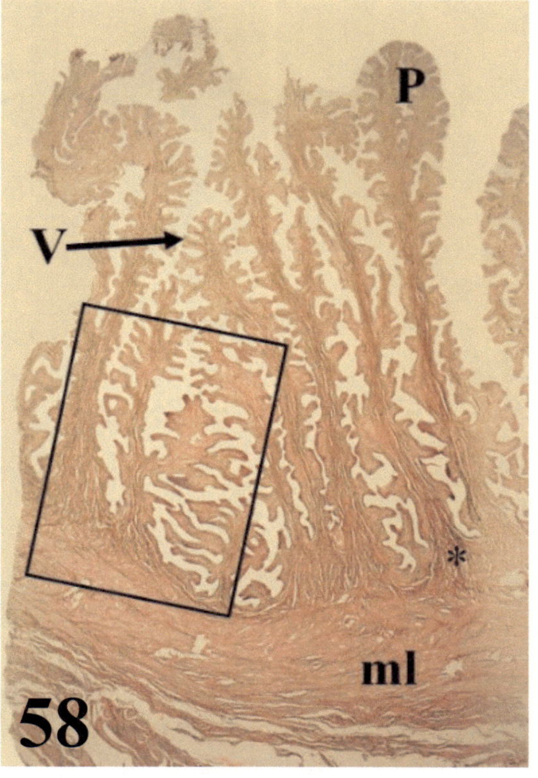

Figure 58.
Panoramic histological section of the vaginal wall with long, branched folds (P) and villi (V). Smooth muscle (ml) with the internal circular arrangement and external plexiform. The first is projected into the submucosa (*) of the folds and their ramifications. Staining: HE, 12.5×.

musculature, which extends from the esophagus to the reticular groove; and in the case of the horse, this striated muscle comprises the proximal two-thirds of the esophagus, but gradually changes to smooth muscle in the distal third [20]. In the ostrich proventriculus, protein digestion begins through the secretion of pepsinogen and hydrochloric acid, given in part by the secretion of poorly developed simple and branched tubular fundic glands from the mucosa. The greatest contribution of hydrochloric acid, apparently, is the product of the tubuloalveolar glands composed of the submucosa [12]. In the ostrich, between the muscular layer and the serosa, we find a thick layer of compact connective tissue of parallel bundles of type I collagen; this tendinous layer surrounds the entire organ. Such a structure was also observed in the abomasum wall of an adult South American camelid, the llama (Lama glama) [18]. In the muscular ostrich stomach, the mucosa is covered by a thick layer with a horny appearance, which corresponds to the cuticle (glycocalyx), similar to that described in the adult rooster [22]. The muscle in the ostrich is similar to the chicken, it has three thick layers of smooth muscle fibers. However, in the chicken, in some sectors, only a layer of circularly arranged smooth muscle fibers can be seen [22].

Throughout the mucosa of the small intestine, long intestinal villi are observed, also described in the chicken. In ruminants, they are short and thick structures [16]. In ruminants and pigs, eosinophilic granule cells called Paneth cells [20] are observed at the bottom of the intestinal crypts. These cells were not observed in ostriches (Bezuidenhout) [23]. The muscularis mucosa in the ostrich duodenum and jejunum

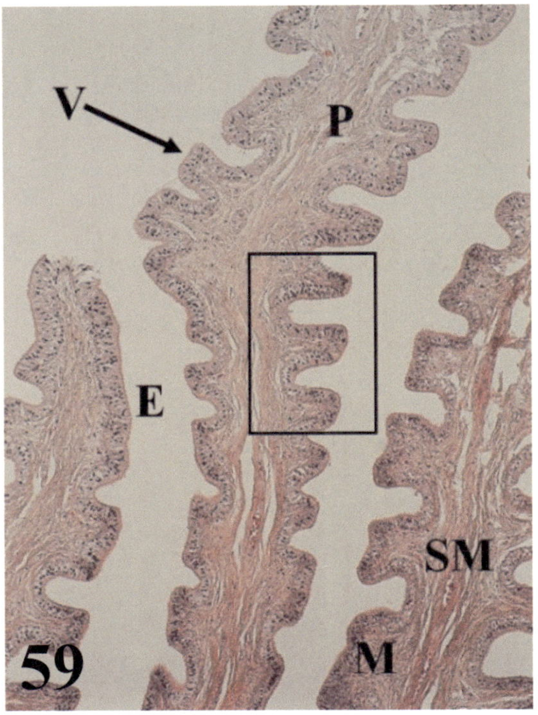

Figure 59.
Higher magnification of the inset of ***Figure 58****. E: mucosal epithelium (M) lining the folds (P) and their villi (V). Submucosa (SM). Staining: H-E, 200×.*

presents two layers of smooth muscle fibers, one internal circular and one external longitudinal, and three layers of muscle fibers are described in the ileum. On the other hand, in the hen, the muscularis mucosa consists of only one layer of smooth muscle fibers arranged longitudinally along the entire segment [22]. The submucosa in the ostrich, as in the chicken, is observed as a thin layer of loose connective tissue, it shows a greater thickness when it contains submucosal nerve plexuses or blood vessels and does not show glands [17, 21]. In contrast, in ruminants and horses, the submucosa shows tubuloalveolar glands, whose excretory ducts cross the muscularis mucosa and open at the bottom of the intestinal crypts. These glands are of the mucous type in ruminants and of the serous type in the horse, which are observed up to the jejunum [20].

At the level of the large intestine, in the cecum of the ostrich a spiral fold is described, formed by a nucleus of mucosa and submucosa, this plays an important role in the absorption of volatile fatty acids and other metabolites produced by the microbial fermentation of cellulose and hemicellulose (Bezuidenhout) [23]. The colon and rectum present longitudinal folds made up of mucosal and submucosal nuclei, as described for other domestic species [20]. The mucosa in the ostrich's large intestine lacks villi and is formed by a simple columnar epithelium, similar to that described in the horse and in the ruminant [20]. The microscopic analysis carried out on the glands and the different segments of the ostrich digestive tract shows the general structure present in the wall of the digestive tract of vertebrates. However, certain differences are observed that may be the product of functional and adaptation changes. This histological analysis is a contribution not only to biology, but can also be very useful,

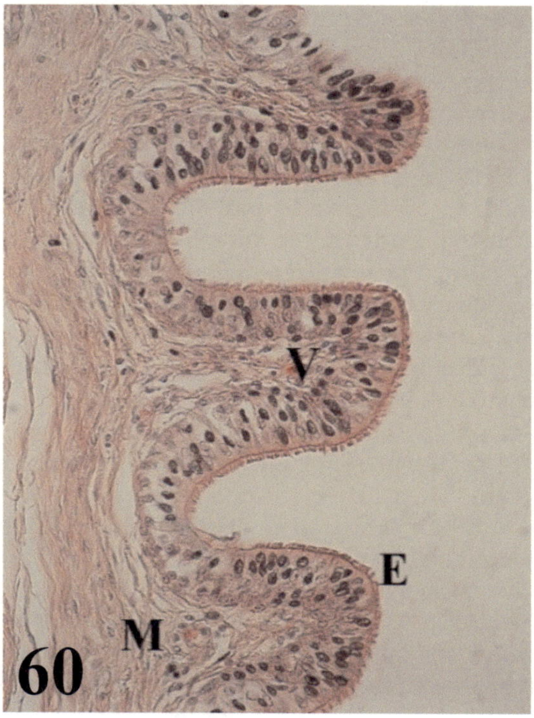

Figure 60.
*Higher magnification of the inset of **Figure 59**. Shows the mucosa (M) with its pseudostratified lining epithelium prismatic ciliated with goblet cells (E) and villi (V). Staining: H-E, 400×.*

especially to better understand the food and nutritional processes during ostrich breeding and management. In addition, they constitute a tool to face the diseases that are produced in them.

The conduction pathway from the larynx to the secondary bronchi is characterized by a mucosa lined by a ciliated prismatic pseudostratified epithelium with goblet cells, interrupted by crypts with simple mucus-secreting tubuloalveolar glands. With the exception of the extrapulmonary primary bronchi, which show this characteristic only in the mucosa of the medial region where the ends of the hyaline cartilage semi-rings of their adventitia converge. The remaining mucosa of these presents a ciliated prismatic pseudostratified epithelium with goblet cells without crypts or simple tubuloalveolar glands with interposed mucus secretion [14]. However, it should be noted that in the mucosa of the air sacs, crypts with simple tubuloalveolar glands of mucus secretion can be observed in some sectors. In these air sacs, the mucosa is reinforced with type I collagen fibers, constituting a tendinous layer [14]. This tendinous layer of type I collagen is observed in the ostrich proventriculus subserosa [13] and in the abomasum subserosa in the llama [18]. The ostrich syrinx is located at the bifurcation of the trachea, it presents three thickenings of the mucosa, the lateral tympaniform membranes, the medial tympaniform membranes that are analogous to the vocal folds of mammals [24], and the central membrane or pessulus, lined by ciliated prismatic pseudostratified epithelium with goblet cells, also observed and described by [25]. However, crypts with simple mucus-secreting tubuloalveolar glands interposed in their epithelium are also observed. In the ostrich pessulus, neither cartilage nor bone tissue is seen [14]. In the chicken, the pessulus presents a skeleton of cartilage or bone tissue [26, 27].

At the intrapulmonary level, the primary intrapulmonary bronchi or mesobronchi in the ostrich are lined by a ciliated prismatic pseudostratified epithelium with goblet cells, interrupted by crypts with simple mucus-secreting tubuloalveolar glands [14]. The hyaline cartilage of its adventitia is not arranged in semi-rings, but in cartilaginous plates that gradually decrease in size until they disappear at the level of the secondary bronchi. Unlike the ostrich, the intrapulmonary primary bronchi in the chicken have a simple ciliated prismatic epithelium, with numerous cells of mucus secretion [27]. The secondary bronchi in the ostrich present a ciliated prismatic pseudostratified epithelium with goblet cells, with deep crypts ending in simple mucus-secreting tubuloalveolar glands. They do not have hyaline cartilage; however, they show bundles of smooth muscle in a circular arrangement [14]. In the case of chickens, they present a simple ciliated prismatic epithelium with mucous cells [27]. The parabronchi in the ostrich present a low simple cuboidal epithelium over a thin layer of connective tissue. Then a layer of smooth muscle in a helical arrangement in both directions [14]. Various dilations or cavities called atria originate from this bronchial mucosa, which conduct air toward the air capillaries, also described by [28]. The parabronchi in the chicken present the same type of epithelium, with the abundant presence of atria [26, 27]. The air capillaries of the ostrich and the chicken are very similar, they present a simple epithelium, where flat cells (type I pneumocytes) and cuboidal cells (type II pneumocytes) are observed, similar to the epithelium of the lung alveoli in mammals [14, 27]. The mucosa of the conducting airways in the ostrich is adapted to generate large amounts of mucus. This mucus production capacity allows foreign particles to be retained, and temperature and humidity regulation of the inspired air. In this way, it protects this extensive mucosal surface from drying out, given the ecological characteristics where this species reproduces, develops, and lives. Through the comparative analysis, it was possible to establish similarities and differences with the chicken, with special emphasis on the mucous layer. Although many of its characteristics remind us of those observed in other bird species, some differences were observed.

The ovarian cortex in the ostrich is covered by a lining epithelium that changes from simple cuboidal to simple ciliated prismatic and in sectors to pseudostratified prismatic, whereas in the chicken it is simple cuboidal [8, 27]. In the ostrich, the follicles show a microscopic structure similar to those of the chicken [27, 29]. However, in the ostrich, follicles in different stages of development show myofibroblasts concentrically arranged in the theca surrounding the granulosa layer. Myofibroblasts are very noticeable in the theca interna of vitellogenic follicles [15].

In the ostrich oviduct, the mucosa of the folds of the infundibulum, magnum, isthmus, uterus, and vagina present a ciliated prismatic pseudostratified epithelium with goblet cells. In the hen, it is the same, except for the infundibulum which presents a simple ciliated prismatic epithelium [27]. In the chorion of the ostrich oviduct, at the level of the infundibulum, simple alveolar and tubuloalveolar glands of mucous secretion are observed. This type of gland is also found in abundance in the chorion of the esophageal and respiratory mucosa of these birds [14, 18]. The chorion of the magnum presents branched tubular glands of serous secretion. The chorion of the isthmus presents simple tubular glands of serous secretion. The chorion of the uterus has branched tubular glands of serous secretion, and the chorion of the vagina lacks glands. In the chicken, it is similar, however, the chorion of the infundibulum does not present glands and that of the vagina presents simple tubular glands of serous secretion [27]. These would contribute to forming the cuticle of the eggshell [30] and act as a sperm reservoir [10, 27]. In the ostrich, it is likely that the labyrinth of folds, ramifications, and villi present in the vagina acts as a reservoir for spermatozoa. The histological study of the reproductive system of the female will be one more tool that will allow for solving reproductive and management problems in ostrich breeding. In addition, it will be very useful to face the diseases that occur in it.

Author details

Julio Luis Erwin Illanes Herrero
Veterinary Medical, Facultad de Medicina, Clínica Alemana Universidad del Desarrollo, Santiago, Chile

*Address all correspondence to: jillanes@udd.cl

References

[1] Deeming DC. Introducción. In: El Avestruz: Biología, producción y sanidad. Zaragoza: Acribia; 2001. pp. 1-11

[2] Camiruaga, M. El Avestruz: Sistema de Producción en Chile. Fundación para la Innovación Agraria, Santiago, Chile, 2004

[3] Camiruaga M, Simoneti C. Avestruces, Sistema digestivo y su alimentación. 2003. Available from: http://www.faif.puc.cl

[4] Bezuidenhout AJ. Anatomía. In: El Avestruz: Biología, producción y sanidad. Zaragoza: Acribia; 2001. pp. 13-50

[5] Angel CR. A review of ratites nutrition. Animal Feed Science and Technology. 1966;**60**:241-246

[6] Huchzermeyer F, W. Patología de Avestruces y otras Ratites. Madrid: Ediciones Mundi-Prensa; 2000

[7] Dyce KM, Sack WO, Wensing CJG. Anatomia Veterinaria. 2a ed. Ciudad de México: McGraw-Hill Interamericana; 2004

[8] Fudge MA. Laboratory Medicine Avian and Exotic Pets. California: Saunders; 2000

[9] Whittow GC. Sturkies Avian Physiology. 5a ed. Hawaii: Academic Press; 2000

[10] Hafez ESE. Reproducción e Inseminación Artificial en Animales. 7th ed. Ciudad de México: McGraw-Hill Interamericana; 2002

[11] Luna L. Manual of Histologic Staining Methods of the Armed Forces Institute of a Pathology. 3rd ed. New York: McGraw-Hill; 1968

[12] Illanes J, Fertilio B, Quijada M, Leyton V, Verdugo F. Histologic description of the annexed glands from the ostrich digestive system Struthio camelus var. domesticus. International Journal of Morphology. 2006;**24**(3): 297-302

[13] Illanes J, Fertilio B, Chamblas M, Leyton V, Verdugo F. Histologic description of the different segments from the ostrich digestive system (Struthio camelus var. domesticus). International Journal of Morphology. 2006;**24**(2):205-214

[14] Illanes J, Leichtle J, Leyton V, Tapia LF, Fertilio B, Castro M. Histologic description of the different segments from the ostrich respiratory system (Struthio camelus var. domesticus). International Journal of Morphology. 2014;**32**(4):1325-1336

[15] Illanes J, Matus F, Leyton V, Venegas F, Fertilio B, Castro M. Histological description of the ostrich female reproductive system (Strutio camelus var. Domesticus). International Journal of Morphology. 2018;**36**(4): 1252-1261

[16] Bacha WJ, Wood LM. Aparato Digestivo. In: Atlas color de Histología Veterinaria. Buenos Aires: Intermédica; 1991. pp. 111-150

[17] McLelland J. Atlas en color de Anatomía de las aves. En aparato digestivo. Buenos Aires: Interamericana; 1992. pp. 53-54

[18] Illanes J, Vigneaux IM, Adaro L, Olivares R, Mendoza J, Riveros V.

Análisis morfológico comparado de la región fúndica entre un cánido (Canis familiaris) y un camélido suramericano (Lama glama). Revista chilena de anatomia. 1994;**12**(2):161-167

[19] Cummings J, Munnell J, Vallenas A. The mucigenous glandular mucosa in the complex stomach of two new-world camelids, the llama and guanaco. Journal of Morphages. 1972;**137**: 71-110

[20] Stinson AW, Calhoun ML. Sistema Digestivo. En: Dieter Dellmann, H. Histología Veterinaria. Zaragoza, Acribia, 1993. pp. 239-256

[21] Dieter Dellmann H, Eurell J. Textbook of Veterinary Histology. 5th ed. Philadelphia: Lippincott Williams & Wilkins; 1998

[22] Delhon GA, Fernández-Surribas J, Gagna G, Lacolla D, von Lawzewitsch I. Aparato digestivo de las aves domésticas. En: Lecciones de histología veterinaria. Hemisferio Sur, Buenos Aires, 1984. pp. 45-111

[23] Cooper R, Mahroze K. Anatomy and physiology of the gastrointestinal tract and growth curves of the ostrich (*Struthio camelus*). Animal Science Journal. 2004;**75**:491-498

[24] Nickel R, Schummer AY, Seiferle E. Anatomy of the Domestic Birds. Berlin: Verlag Paul Parey; 1977. p. 65

[25] Yildiz H, Bahadir A, Akkoç A. A study on the morphological structure of syrinx in ostriches (Struthio camelus). Anatomia, Histologia, Embryologia. 2003;**32**(3):187-191

[26] Sisson S, Grossman JD. Anatomía de los animales domésticos. Tomo II. 5a ed. México D. F: Masson; 2000

[27] Bacha WJ, Bacha LM. Atlas Color de Histología Veterinaria. 2a ed. Buenos Aires: Intermédica; 2001

[28] Maina JN, Nathaniel C. A qualitative and quantitative study of the lung of an ostrich, *Struthio camelus*. Journal of Experimental Biology. 2001;**204**(Pt. 13): 2313-2330

[29] Kimaro WH. An immunohistochemical and ultrastructural study of the ovary of the immature ostrich (*Struthio camelus*) [M.Sc. thesis]. Pretoria: Department of Anatomy and Physiology, Faculty of Veterinary Science, University of Pretoria; 2005. Available from: https://repository.up.ac.za/bitstream/handle/2263/23731/Complete.pdf?sequence=3

[30] Fujii S. Histological and histochemical studies on the oviduct of the domestic fowl with special reference to the region of uterovaginal juncture. Archivum Histologicum Japonicum. 1963;**23**:447-459

Chapter 9

Distribution of House Sparrows, *Passer domesticus indicus*, in Coimbatore District, Tamil Nadu, India

Mahalakshmi Sundarapandian and Priya Raman

Abstract

The Indian house sparrow, *Passer domesticus indicus*, is a small bird that is common in India, found in close contact with humans, instead of forests. Over recent years, according to the study, the house sparrow population has been on the decline in many Asian countries, which is quite evident in India. To understand the reason for their decline a study was conducted in Coimbatore, Tamil Nadu, India, during 2007–2008. The population of the house sparrows was counted at five sampling sites: agricultural area, marketplace, bus stand, residential area, and industrial area. The industrial area showed a decline in the population compared to other areas. The study was conducted again after 12 years at the same sampling site during the pandemic. It was observed an increase of 128% in the agricultural area, 233% in the marketplace, 369% in the bus stand area, 131% in the residential area, and 214% in the industrial area. It was interesting to notice a pronounced increase in all the study sites, which is slightly deviating from the other studies that reported their decline. The increase in numbers may be due to the fewer working men and less usage of the equipment during the pandemic.

Keywords: house sparrows, *Passer domesticus*, agricultural, residential, industrial, bus stand, market place, COVID-19 pandemic

1. Introduction

The Indian house sparrow, *Passer domesticus indicus*, is one of the commonest birds in India. The species is native to and resides in urban and rural areas across India, Burma, Africa, Britain, and Northern Siberia [1]. The house sparrows are exceptional 'invaders' [2] of the tropics and have 'followed humans worldwide'. It is a non-migratory species and is closely tied to human activity. The species breed in small colonies or in loose groups [3, 4]. House sparrows are commonly found around human dwellings and feed on grains and insects. House sparrows are opportunistic feeders and adapt to a variety of food sources. In the 1920's the house sparrow was regarded as a very serious pest of the growing crops. Later it was observed that house sparrows effectively control many pests of crops. It is also said these birds fly high during the rainy season predicting rainfall. In recent years, there have been reports of the house sparrow population declining throughout Europe [5, 6].

The small-sized bird is known for its distinctive brown-gray plumage and its familiar chirping call. It is monogamous [4] with a small stout conical beak. The bird has a typical length of 16 cm (ranging from 14 to 18 cm), and its mass ranges from 24 to 39.5 g. Males are usually larger than the females. House sparrows also engage in social activities such as dust or water bathing and social singing in which birds call together in shrub bushes, or trees [7]. In recent years, the house sparrow population has been declining in many Asian countries, which is quite evident in India.

2. Textile capital city of Tamilnadu

Coimbatore is a city in the South Indian state of Tamil Nadu and is the second largest city in Tamil Nadu [8]. Coimbatore villages are peaceful and scenic with agricultural and artisanal traditions. Coimbatore is one of the most industrialized cities in Tamil Nadu, known as the textile capital of South India or the Manchester of the South. The city is situated on the banks of the river Noyyal. It is regarded as one of the most urbanized and polluted cities in Tamil Nadu [8]. The town is rapidly deteriorating into environmental decay and degeneration. Rapid industrialization and unlimited urbanization have created environmental disasters. Studies of the urban bird species in India are still in their infancy. There is a dearth of information regarding their distribution, population, dynamics, and threats.

House sparrows have been abundant in the Coimbatore region in the last two or three decades. Of late, their population has been declining at an alarming rate in this region [9]. These birds are an integral part of the ecosystem, and their loss can have cascading effects on other species. It is important to conduct research to understand the reasons behind the decline in the population of Indian house sparrows. This can help in identifying effective conservation strategies that can address the root causes of the problem. As no valid data on their decline are available, a study was conducted to know the incidence of house sparrows in different locations in Coimbatore.

3. Taxonomy and geographic range

Kingdom	Animalia
Phylum	Chordata
Class	Aves
Order	Passeriformes
Family	Passeridae
Genus	*Passer*
Species	*domesticus*
Subspecies	*indicus*

The house sparrows are culturally prominent and found in most parts of the world. This small bird belongs to the order of the perching birds and sparrow family. The genus *Passer* contains about 25 species according to the Handbook of Birds of the World. There are more than 12 recognized subspecies [10]. These subspecies vary in their distribution and characteristics, reflecting regional adaptations and genetic diversity. Few recognized subspecies

1. *Passer domesticus domesticus* (found in Europe, Asia and North Africa)

2. *Passer domesticus indicus* (found in the Indian subcontinent)

3. *Passer domesticus bactrianus* (found in Central Asia)

4. *Passer domesticus tingitanus* (found in Morocco and parts of North Africa)

5. *Passer domesticus niloticus* (found in parts of Sudan and Ethiopia)

6. *Passer domesticus balearoibericus* (found in the Balearic Islands of Spain)

The *Passer domesticus* was described as a species and *Passer domesticus indicus* was considered a distinct subspecies by many ornithologists during the 19th century [11]. *Passer domesticus indicus* have white cheeks and is widely distributed in the Indian subcontinent and is also found in Southern Israel and most of the Arabian Peninsula east to southern Iran, India, and Sri Lanka [12].

4. The study protocol

The study areas were chosen based on the types of habitats in different directions. The population trends of the house sparrows were counted at five sampling sites (**Figure 1**), namely, agricultural area with farmlands (Kovanoor), densely populated marketplace (Periyanaickenpalayam), highly vehicular traffic area – bus stand (Veerapandi pirivu), moderately populated and moderately vehicular traffic residential area (Press colony) and highly polluted industrial area (Shanthimedu).

The counting of birds was carried out in the morning from 7 to 10 AM when they are most active and conspicuous [13]. No single survey can provide all the necessary data for every research question [14]. Surveys were, therefore, conducted once a fortnight in the identified locations [13] from June 2007 to February 2008. After a long period, the same places were selected and observed during the lockdown period from June 2020 to February 2021. Regular field observation was also made on the nest, nesting sites, feeding habitats, and food sources [13]. Recordings were also made season-wise. The number of sparrows was conducted by field methods, recording methods, and distance estimates. In this study, the number of sparrows was counted by the point count method. The census method involving repeated counts in the same place in successive months is likely to provide an accurate picture of the population status [15, 16].

5. Results

The observations made on the abundance of *Passer domesticus indicus* populations in different regions are given in **Table 1**. The data reveals that population declines have occurred in both urban and rural areas, but more in the former [15]. This should be a serious public concern [17]. The agricultural area selected for the study

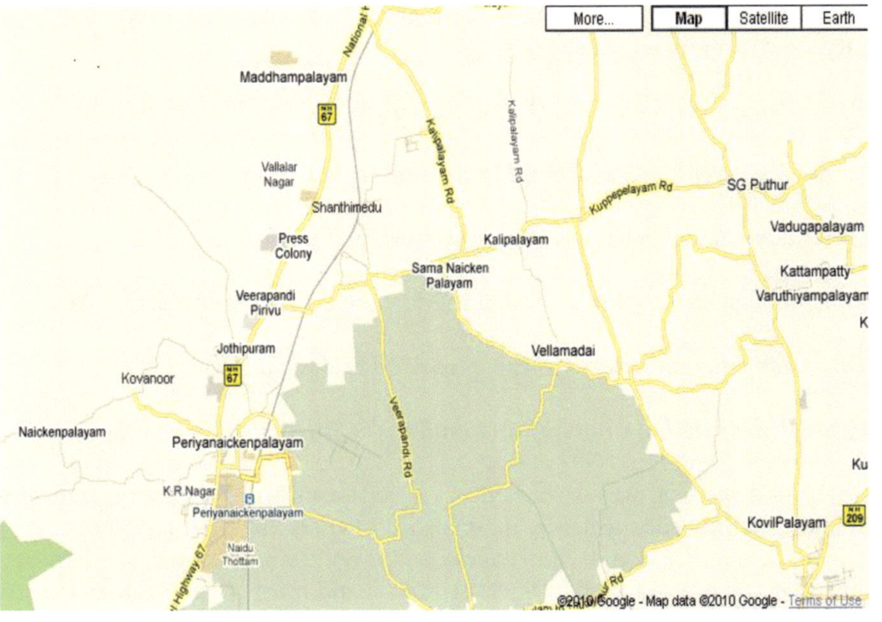

Figure 1.
Study site.

Months	Agriculture	Market	Bus Stand	Residential	Industrial
June	25	10	4	17	7
July	27	7	2	15	5
August	20	9	3	14	6
September	33	15	5	23	3
October	35	12	4	27	2
November	31	18	3	30	5
December	39	19	4	34	4
January	32	23	5	30	2
February	35	15	3	29	7
Total	277	128	33	219	41
Status	I	III	V	II	IV

Table 1.
June 2007 to February 2008.

is Kovanoor. Most of the population is engaged in farming activities. It is situated near the Noyyal River and several smaller streams, which provide water for irrigation and contribute to the area. The highest number of 277 house sparrows were observed in the agricultural area. Moderate numbers of 219 sparrows were recorded in the residential (Press colony). They are adapted to living in close proximity to humans and are often found in gardens, parks in residential areas.

One hundred twenty eight sparrows were recorded in the market area (Periyanaickenpalayam). In market areas, they may be attracted to food stalls, where they can scavenge leftover food and crumbs. They may also be attracted to areas where grains or other food items are stored or sold. The population has undergone a significant decline in the number of 41 in the industrial area (Shanthimedu). While house sparrows can adapt to living in industrial areas, it is important to be aware of the potential hazards associated with industrial activity. Thirty three were recorded in the bus stand area, Veerapandi Pirivu. House sparrows are not commonly found in bus stand areas as these environments are typically noisy and heavily trafficked by people and vehicles. Additionally, bus stand areas tend to have a limited amount of vegetation and green spaces, which can make it difficult for birds to find suitable nesting sites and foraging opportunities [18].

Conservation efforts in all these environments can help to promote the diversity of bird populations and support the health of the ecosystem. The population of house sparrows was seen to be more pronounced in the winter period [19] (December, January, February) than in the spring (June, July, August) and autumn (September, October, November) period. This increase in the number of birds in winter may be due to successive breeding of the sparrows. Certainly, house sparrows can continue to coexist with humans and flourish where they live.

It was observed that the population of these birds has increased during the pandemic, shown in **Table 2**. The birds showed an increased population of 78 in the agricultural area, 171 in the marketplace, 89 in the bus stand area, 70 in the residential area, and 47 in the industrial area. This may be due to the lesser vehicular usage. In the industrial area, the population was pronounced high compared to the other months when the study was conducted particularly during the breeding season. The percentage increase in the house sparrow is given in **Table 3**. This shows that the population has increased in the bus stand area by 369% followed by the marketplace with 233%, industrial area by 214%, residential area by 131%, and agricultural area by 128%. All the other study areas showed more birds compared to the earlier study. The COVID–19 pandemic has primarily affected humans with social and economic impacts worldwide. This has changed human behavior like less travel and outdoor activities during lockdown. This may also play a role in the increase in the population.

Months	Agriculture	Market	Bus Stand	Residential	Industrial
June	29	20	9	20	8
July	27	22	7	21	7
August	25	27	10	26	9
September	35	31	14	29	9
October	39	37	10	30	7
November	45	34	16	33	8
December	53	38	15	40	15
January	50	43	19	48	12
February	52	47	22	42	13
Total	355	299	122	289	88
Status	I	II	IV	III	V

Table 2.
June 2020 to February 2021.

Sampling site	Number increase	% increase	Status
Agricultural	78	128	V
Market	171	233	II
Bus Stand	89	369	I
Residential	70	131	IV
Industrial	47	214	III

Table 3.
Percentage of increase in the house sparrows.

6. Discussion

The decrease in the population of house sparrows may be due to the unavailability of plenty of food such as grains, insects in the vegetables (caterpillars), and suitable nesting sites [20] in the agricultural area [13, 15]. It has been reported that house sparrows usually build their nests in the crevices of thatched roofs of old houses, electric pipelines, ventilation holes, and in the space available on the electricity meters [21]. Planting hedgerows and other native vegetation to provide food and shelter for birds and creating nesting sites [22] for birds may improve their population. Farmers can control the spread of non-native animals such as feral cats or European starlings, which can compete with house sparrows for food and nesting sites. It is also known that sparrows in small groups rest on hanging electric wires in city areas [13]. A similar pattern was observed in the present study also, shown in **Figure 2**.

The decline in the number in residential areas is due to the presence of cell phone towers [23]. Gardens are increasingly popular in Coimbatore. They are often treated as external rooms, where bird feeding is done with good quantity and a range of foodstuffs [24]. A group of house sparrows chirping in the evening was observed on a shrub (**Figure 3**) and in garden landscapes (**Figure 4**). The dismantling of tiled houses and construction of concrete buildings has become a problem to house sparrows for their living and breeding. Modern buildings are generally without roof spaces for nesting [25]. Roof repairs and loft conversions are more prevalent in wealthier areas. Even when the nesting sites are available crows frequently destroy the chicks by either pulling them out of the nest or capturing them when they take their first flight [26–28].

Many buildings and landowners are leasing their rooftops and nearby garden land to telecommunication companies. The owners may not be aware of the hazardous effect. This may also be a reason for their general population decline. One way to provide nesting sites for house sparrows is by installing birdhouses or nesting boxes in suitable locations. These can be designed to mimic natural nesting sites, such as cavities in trees or buildings, and can be placed in sheltered areas around the market (**Figures 5–12**) [29].

Studies on food supplementation experiments have shown that the decline could be related to a reduction in winter food supplies [15, 30]. The moderate number of populations observed in the market area is due to the reduction of food availability. Habitat quality is known to have a major influence on sparrow populations with the availability of food sources [31–33].

In industrial areas, they may feed on scraps of human food found in dumpsters or other waste areas. Increased human activity and vehicle traffic in these areas can

Figure 2.
House sparrows resting on electric wires.

Figure 3.
House sparrows chirping in the evening.

pose a risk to birds. To minimize the impact of noise and other forms of disturbances on house sparrows, it is important to implement measures to reduce noise levels and create suitable habitats that are less susceptible to disturbances.

Figure 4.
House sparrows in the garden landscape.

Figure 5.
Artificial nests mimicking the natural nests. [Hindustan Tiimes, march 21, 2022].

It is important to keep these feeding stations clean and well-maintained to prevent the spread of disease. Minimizing human activity around bird habitats during the breeding season can help to reduce disturbance and promote successful reproduction. Monitoring bird populations is important to ensure that conservation efforts are effective and that bird populations remain healthy and thriving. It is important to work with local authorities and organizations to develop bird-friendly policies

Figure 6.
Artificial nests mimicking the natural nests. [birdbgone.com].

Figure 7.
Artificial nests mimicking the natural nests. [wwwyoutube.com].

and practices in industrial areas, such as guidelines for the management of industrial waste and the use of bird-friendly materials and designs in industrial infrastructure.

In marketplace, there were fewer house sparrows recorded due to the reduction in the number of domestic animals. Due to this reduction, fewer amount of food resources like spilled grains as both feed and dung are available to the house sparrows [34]. The causes for the decline of the house sparrows may also include a reduction of food supplies, possibly from the reduced refuse along the roads, more efficient

Figure 8.
Artificial nests mimicking the natural nests. [www.sialis.org].

Figure 9.
Artificial nests mimicking the natural nests. [amazon.in].

street cleaning, a reduction of weedy areas in both private and public gardens [10], and a reduction of private small scale poultry farming practices in rural areas [14]. The existence of few houses with limited availability of nesting sites and food availability is also a reason for their lesser population. By creating green spaces, and a more hospitable environment for house sparrows we can help to ensure that these important habitats continue to support diverse bird populations [35]. Reducing pollution can help improve air and water quality while promoting the coexistence of birds and humans can help to promote a sense of community and connection to nature.

Figure 10.
Artificial nests mimicking the natural nests. [© Oscar Gutierrez].

Figure 11.
Artificial nests mimicking the natural nests. [times of India, march 20, 2020].

7. Review of the global and Indian situation

The House sparrow population has been dwindling in many parts of the world, and India is no exception. Noyyal River is a vital ecosystem that supports a wide range of plant and animal species, including several species of birds. House sparrows may be attracted to areas where there is vegetation cover, such as grassy areas and scrublands. They may also be attracted to areas with human-made structures, such as buildings or warehouses, where they can build their nests in crevices or holes. The house sparrows have adapted to living near humans, their populations have been

Figure 12.
Artificial nests mimicking the natural nests. [odishabytes.com, march 20, 2021].

declining in recent years, and conservation efforts are needed to help support their populations. There is thus a serious threat to the survival of house sparrows. Unless they adapt to the changes, house sparrows may migrate or gradually disappear. House sparrows are known to thrive in rural and agricultural environments with a mix of green spaces, open areas, and buildings with suitable nesting sites.

Species, by their very nature, have specific ecological requirements. They may appear to share the same habitat as many others, but each has a different specific niche [36]. Creating bird-friendly habitats, providing supplemental food and water sources for birds, particularly during drought or when natural food sources are scarce. Pesticide use can be harmful to birds and other wildlife, so it is important for farmers and residents in the area to minimize the use of harmful chemicals [37]. Protecting natural water sources and ensuring that agricultural practices are sustainable and environmentally friendly can help maintain and enhance the biodiversity of the area. Farmers and other landowners can also work to control the spread of invasive plant species, which can displace native vegetation and reduce available habitat for birds. Habitats must, however, meet the needs of the species that depend on them [28]. Adoption of sustainable practices and technologies [3] that reduce the environmental impact of industrial activity i.e., by use of renewable energy sources such as solar power.

In addition to habitat loss and pesticide use [37], other factors that contribute to the decline in the population of Indian house sparrows include air pollution, electromagnetic radiation [5, 23], and the lack of nesting sites. As urbanization and development continue to expand in India, it is important to take steps to conserve the natural habitats of these birds and promote their survival.

Conservation is not just about avoiding extinctions but also restoring or recovering species populations to secure levels [36]. The rapid decline of house sparrows is a serious concern because,

- If such a human-associated bird species is suddenly declining, the situation seems to be bleak. The future looks even bleaker for other species that are more sensitive to human habitat change and require more pristine habitats.

- The implication of the decline in sparrow populations is also reflected in the living conditions in the human environment, particularly in the cities.

Conserving and restoring habitats on the landscape scale is vital to nature conservation, especially to make biodiversity robust to environmental change. Apart from conservation efforts, research on the biology, behavior, and habitat requirements of Indian house sparrows are important to understand the reasons behind their decline and to develop effective conservation strategies [38]. Raising awareness through educational initiatives, such as workshops or public outreach programs, as well as through media campaigns about the importance of conserving house sparrow populations can help to promote positive attitudes and behaviors towards bird conservation.

Many government and Non-Governmental Organizations (NGOs) in India are working together to conserve house sparrows. The Nature Forever Society, a Mumbai-based organization, has launched a nationwide campaign called "Save Our Sparrows" to promote the conservation of these birds. The organization motivates people to fabricate suitable habitats for sparrows, providing nest boxes, and bird feeders and promoting awareness about the importance of conservation.

Rural areas in and around Coimbatore are facing a diversity of environmental challenges, like climate change, water scarcity, and soil degradation. Addressing these challenges will require a collaborative effort from the local community, as well as support from the government and other organizations. The situation in Coimbatore is thus alarming. Immediate conservation measures are, therefore, required to be carried out in the coming years. Conservation plans and status assessments at the national level should be implemented to check the decline of sparrow populations. By working together, it may be possible to ensure that places remain sustainable and lively for future generations.

8. Conclusions

In conclusion, monitoring bird populations is important to make sure that conservation efforts are effective, and that populations remain healthy and flourishing. This can be done through citizen science programs, or other monitoring initiatives, which can help to identify trends in bird populations and track the effectiveness of conservation efforts. Trailing broader environmental and social goals can help to promote the coexistence of birds and humans in these environments. The *Passer domesticus indicus* is an important bird species that plays a significant role in maintaining the ecological balance in urban and rural areas of India. One must strive hard to conserve this species by promoting awareness, reducing pesticide use, and promoting suitable nesting sites and food sources for the birds. Efforts from individuals, organizations, and the government are needed to promote their survival and ensure that future generations can enjoy the presence of these charming and important birds. The decline of house sparrows is not unique but has been observed in other parts of the world as well. Therefore, international collaboration and research can help develop effective strategies for the conservation of these subspecies globally.

Acknowledgements

We thank the staff members of the Department of Zoology, Kongunadu Arts and Science College for their help and support.

Author details

Mahalakshmi Sundarapandian[1*] and Priya Raman[2]

1 School of Life Sciences (Ooty Campus), JSS Academy of Higher Education and Research, Ooty, The Nilgiris, Tamil Nadu, India

2 School of Agricultural Sciences, Amrita Vidya Vishwapeetham, Coimbatore, India

*Address all correspondence to: mahasundar@jssuni.edu.in

References

[1] Harrison J. A Field Guide to the Birds of Sri Lanka. UK: Oxford University Press; 1999. p. 193

[2] MacGregor-Fors I, Quesada J, Lee JG-H, Yeh PJ. Space invaders: House sparrow densities along three urban–agricultural landscapes. Avian Conservation and Ecology. 2017;**12**(2)

[3] Suraj Prabhakar Naik. Lake Proceedings: Conference on Conservation and Sustainable Management of Ecologically Sensitive Regions in Western Ghats, Sahyadhri Conservation Series 65. Bangalore: Indian Institute of Science; 2016

[4] Veiga JP. Why are house sparrows predominantly monogamous? Animal Behaviour. 1992;**43**(3):361-370

[5] Balmori A, Hallberg O. The urban decline of the house sparrows (*Passer domesticus*): A possible link with electromagnetic radiation. Electromagnetic Biology and Medicine. 2007;**26**(2):141-151

[6] Summer-Smith JD. Current status of house sparrow in Britain. British Wildlife. 1999;**10**:381-386

[7] Leasure DR, Kannan R, James DA. House sparrows associated with reduced cliff swallow nesting success, The Wilson Journal of Ornithology. 2010;**122**(1):135-138

[8] Available from: www.coimbatore.tn.nic.in

[9] Narayanappa Y, Gautam A, Mahobiya K, Singh A. A pilot study on the occurrence and portable factors influencing the population decline of house sparrow (*Passer domesticus*) along an urbanization gradient in Coimbatore district, India. Biodiversitas. 2022;**23**(8):3884-3889

[10] Summers-Smith JD. Densities of house sparrows in different urban habitats in a small town in NE England. Journal of Yamashina Institute of Ornithology. 2009;**40**:117-124

[11] Vaurie C, Koelz W. Notes on some Ploceidae from Western Asia. New York: American Museum Novitates; 1949

[12] Jardine W, Selby PJ. Illustrations of Ornithology. 1st ed. Edinburg: WH Lizars; 1831

[13] Rajashekar S, Venkatesha MG. Occurrence of house sparrow, *Passer domesticus indicus* in and around Bangalore. Current Science. 2008;**94**(4):446-449

[14] Turner WR. Citywide biological monitoring as a tool for ecology and conservation in urban landscapes: The case of the Tucson bird count. Landscape and Urban Planning. 2003;**65**(3):149-166

[15] Hole DG, Whittingham MJ, Bradbury RB, Anderson GQA, Lee PLM, Wilson JD, et al. Widespread local house sparrow extinctions-agricultural intensification is blamed for the plummeting population of these birds. Nature. 2002;**418**:931-932

[16] Brichetti P, Rubolini D, Galeotti P, Fasola M. Recent declines in urban Italian Saprrow *passer* (*domesticus*) Italine populations in western Italy. Ibis. 2008;**150**:177-181

[17] Robinson RA, Siriwardena GM, Crick HQP. Size and trends of house

sparrow (*Passer domesticus*) population in Great Britain. Ibis. 2005;**147**(3):552-562

[18] Shaw LM, Chamberlain D. The house sparrow *Passer domesticus* in urban areas: Reviewing a possible link between post decline distribution and human socioeconomic status. Journal of Ornithology. 2008;**149**:293-299. DOI: 10.1007/s10336-008-0285-y

[19] Berigan LA, Greig EI, Bonter DN. Urban house sparrow (*Passer domesticus*) populations decline in North America. The Wilson Journal of Ornithology. 2020;**132**(2):248-258

[20] Maxmellion PME, Rajendran P. A study on the nesting site preference and adaptability of house sparrow to various locations in the rural, suburban and urban gradients. Journal of Science and Technology. 2021;**6**(5):89-98

[21] Ali S. The Book of Indian Birds. 6th ed. New Delhi: Oxford University Press; 1996

[22] Choudhary S. NPS Chauhan, Rajiv Kalsi, finding clues in cities: Nest site selection by house sparrows across a varied habitat matrix in Delhi, India. Current Science. 2020;**118**(8):1304-1309

[23] Everaert J, Bauwens D. A possible effect of electromagnetic radiation from Mobile phone base stations on the number of breeding house sparrows (*Passer domesticus*). Electromagnetic Biology and Medicine. 2007;**26**(3):63-72

[24] Cannon A. The Garden Birdwatch Handbook. 2nd ed. Thetford, UK: British Trust for Ornithology; 2000

[25] Edgar Bernat Ponce. Jose Antonio Gil Delgado, Daniel Guijarro, factors affecting the abundance of house sparrows *Passer domesticus* in urban areas of southeast of Spain. Bird Study. 2018;**65**(3):404-416

[26] Ranjit Daniels RJ. Can we save the sparrow? Current Science. 2008;**95**(11):1527-1528

[27] Wotton SR, Field R, Langston RHW, Gibbons DW. Homes for birds: The use of houses for nesting by birds in the UK. British Birds. 2002;**95**(11):586-592

[28] Pandian M. Habitats of house sparrow *Passer domesticus* (Linnaeus 1758) in Rameshwaram Island, Tamil Nadu, India. Journal of Threatened Taxa. 2023;**15**(2):22586-22596

[29] Internet source – Google photos. Available from: https://www.hindustantimes.com/cities/delhi-news/world-sparrow-day-making-nests-from-scrap-can-attract-the-bird-to-urban-areas-101647836173823.html; https://www.birdbgone.com/blog/how-to-spare-yourself-from-sparrow-troubles; https://www.youtube.com/watch?v=sCXHROyUVhc; https://www.sialis.org/nestshosp.htm; https://www.amazon.in/Nature-Forever-House-Shaped-Sparrow/dp/B00JWIY9IK; https://www.homesandgardens.com/gardens/bird-house-ideas; https://timesofindia.indiatimes.com/city/pune/house-sparrows-take-to-artificial-nest-boxesfour-years-of-conservation-shows-success/articleshow/74720499.cms?from=mdr; https://odishabytes.com/save-house-sparrow-campaign-sees-birds-10-time-population-growth-in-odishas-gopalpur/

[30] MacLeold R, Barnett P, Clark J. Will cress well, mass dependent predation risk as a mechanism for house sparrow declines? Biology Letters. 2005;**2**(1):43-46

[31] Rana DB, Idris LM. Food habits of the house sparrow *Passer domesticus*

indicus in an arid environment. Japanese Journal of Ornithology. 1987;**35**:125-128

[32] Khera N. Arkaja das, Saumya Srivasatava, Siddharth Jain, habitat wise distribution of the house sparrow (*Passer domesticus*) in Delhi, India. Urban Ecosystem. 2010;**13**:147-154

[33] Vaclav R, Hoi BD. Food supplementation affects extrapair paternity in house sparrows. Behavioural Ecology. 2003;**14**:730-735

[34] Laet JD, Summers-Smith JD. The status of the urban house sparrow *Passer domesticus* in North-Western Europe: A review. Journal of Ornithology. 2007;**148**(2):s275-s278

[35] Wilkinson N. Factors influencing the smaller-scale distribution of house sparrows *Passer domesticus* in a suburban environment. Bird Study. 2006;**53**:39-46

[36] Available from: www.rspb.org.uk

[37] Douglas R. Leasure, the house sparrow *Passer domesticus* decline – Conservation tools emerge from a contrasting north American perspective. Indian BIRDS. 2013;**8**(1):22-23

[38] Klok C, Holtkamp R, van Apeldoorn R, Visser ME, Lia Hemerik S. Analysing population numbers of the house sparrow in the Netherlands with a matrix model and suggestions for conservation measure. Acta Biotheoretica. 2006;**54**:161-178